Armored Vehicles and Units
of the German Order Police (Ordnungspolizei)
1936-1945

Werner Regenberg

A Hotchkiss tank coming right at the camera. The maker's name on the bow existed in several variations. The track aprons of this vehicle were lengthened to the front after manufacture. (PFA)

Armored Vehicles and Units
of the German Order Police (Ordnungspolizei)
1936-1945

Werner Regenberg

Schiffer Military History
Atglen, PA

Shielded by a tank, police riflemen advance on a partisan-occupied village in the Olvidino area in September 1942. (DIZ)

Cover picture: Drawing by Heinz Rode, Berlin
Translated from the German by Ed Force
Book design by Ian Robertson.
Copyright © 2002 by Schiffer Publishing.
Library of Congress Catalog Number: 2001096849

This book was originally published under the title, *Panzerfahrzeuge und Panzereinheiten der Ordnungspolizei 1936-1945* by Podzun-Pallas.

Printed in China.
ISBN: 0-7643-1555-2

We are always looking for people to write books on new and related subjects. If you have an idea for a book, please contact us at the address below.

Published by Schiffer Publishing Ltd.
4880 Lower Valley Road
Atglen, PA 19310
Phone: (610) 593-1777
FAX: (610) 593-2002
E-mail: Schifferbk@aol.com.
Visit our web site at: www.schifferbooks.com
Please write for a free catalog.
This book may be purchased from the publisher.
Please include $3.95 postage.
Try your bookstore first.

In Europe, Schiffer books are distributed by:
Bushwood Books
6 Marksbury Ave.
Kew Gardens
Surrey TW9 4JF
England
Phone: 44 (0)208 392-8585
FAX: 44 (0)208 392-9876
E-mail: Bushwd@aol.com.
Free postage in the UK. Europe: air mail at cost.
Try your bookstore first.

CONTENTS

FOREWORD

This work had its origins in the research that I have conducted since the early 1980s on the Wehrmacht and Waffen-SS units equipped with armored vehicles captured from various nations. During my research I also came upon armored troops of the police which were also completely or partially equipped with captured vehicles.

The published literature on the subject of the police in the Third Reich deals overwhelmingly with the Gestapo, action groups of the Security Police and the SD, as well as the inhuman deeds of the police battalions in the occupied Eastern areas. Aside from the work of Neufeldt, Huck and Tessin, published in 1957, there is very little published material on the organizational history of the police, especially the Ordnungspolizei, and even less on their equipment and its use.

Thus even in the literature, no precise data on the formation, equipment and use of armored units of the Ordnungspolizei can be found, which provided the impetus for more thorough research on this particular subject.

This research has not been simple, as records, particularly the war diaries of the officials and commanders of the Ordnungspolizei, as well as those of the troop units, were either destroyed or lost at the war's end. Thus one must turn almost exclusively to secondary sources, files of the homeland service offices, the Wehrmacht, the Waffen-SS, private archives and collections. The printed sources could be added to in part by talking with former members of the Ordnungspolizei's armored troops. This oral history, though, has a weakness, as memories fade, and get mixed with what had been heard later, after forty to fifty years. On the other hand, photographic documentations are genuine, but require a great deal of time to locate in archives, collections and inherited materials. Through knowledge of the development of uniforms and equipment, undated photos can be included successfully, and at times even the localities could be determined.

Difficulties arose in preparing this book, as to date incomplete data on the history of the Ordnungspolizei, its structure and action, is available. In the framework of this study, only little can be said of the organized units and command structure, and sometimes this information is missing completely.

This book is an attempt to describe a special service arm of the Ordnungspolizei, its equipment, organization and action. Despite all our efforts, no complete history of the Ordnungspolizei's armored troops could be assembled. In some units or actions, sources and/or pictures are lacking. But as an overview of all the described units, a fairly complete picture of this service arm emerges.

The language and choice of words used in the sources was partially retained, even when it is somewhat colored by propaganda or used for concealment. For example, according to Himmler's orders, partisans were to be called "bandits" or "gangs" from mid-1942 on. "Pacification" meant putting down any resistance, sometimes to the point of annihilating the populations of entire villages and areas. "Conscription" refers to the recruiting of forced laborers. Larger actions against partisans were called "undertakings" and given code names such as "Hamburg" or "Easter Bunny". Place names occur in the original sources spelled in various ways, and are repeated thus here because, particularly in the case of smaller places, it cannot be ruled out that they may refer to different places.

My thanks go out to all former members of the Ordnungspolizei or their families who supported the compilation of this book with personal accounts, documents and photos. To represent them I would like to thank the former District Officer of the Schutzpolizei, Johannes Fritz, who died in 1998. I would also like to thank the staffs of all the archives and collections that I visited, inside and outside Germany, very heartily for their active assistance in my search for sources, even when success was ruled out by a lack of available materials.

My thanks also go out to the many friends and acquaintances who have dedicated themselves, as I have, to researching areas of military and police history, and who have encouraged and advised me over many years. From among them I might name Hilary Doyle, Thomas L. Jentz and particularly Karlheinz Muench.

Special thanks go to my dear wife, who has made it possible for me to compile this book through her great understanding of my research work and her active assistance, such as critically examining the manuscript. In the future I shall continue to work on filling existing holes in the information on the Ordnungspolizei's armored troops. All who read this book and might be able to correct errors or contribute photos or written material to complete it are urged to get in touch with me.

Dr. Werner Regenberg
Nussloch, 1999

CHAPTER I
INTRODUCTION

Armored Vehicles for the Police

In the establishment of closed (garrisoned) police units in the provinces of the German Reich shortly after World War I, armored road vehicles were provided for the police for the first time in German history. They still made do with armored vehicles of the old army, or copies of them, plus trucks with makeshift (protective) armor. Armored vehicles saw action in large numbers in World War I for the first time in military history, and had proved themselves at the fronts. The disturbances in postwar Germany in the early 1920s also showed the suitability of these vehicles for the police. Thus the competent authorities stressed again and again that the protected truck was the most suitable means of putting down internal unrest. In most cases the morale effect at the appearance of these vehicles was enough to scatter the crowds of people and make use of weapons unnecessary. It was also suggested that, for example, in Berlin only such armored vehicles could respond quickly to protect threatened government and other buildings. The Interallied Military Control Commission (IMKK) requested, in its note of June 22, 1920, the disarming and disbanding of the security police that had existed until then, and was regarded as a military organization not allowed by the Treaty of Versailles, by September 22, 1920. This security police was already equipped not only with tanks, but also with other heavy military weapons, and even flame-throwers. With the disbanding of the security police, a different "police organization", the Ordnungspolizei, could be enlarged from 92,000 to 150,000 men.

The IMKK governed not only the organization but also the arming of the "new" Ordnungspolizei. Accordingly, the 150,000 men of the Ordnungspolizei were allowed these weapons: one simple weapon per man, a pistol and hand grenades, one rifle or carbine per three members, one machine pistol per 20 members and one armored car without tracks per 1000 men, armed with one machine gun (later two machine guns were allowed).

Thus the number of armored cars for the country's police was set at 150 vehicles. By taking over the armored cars of the former security police and the transitional army (the Reichswehr was allowed only 105 armored but unarmed personnel carriers by the IMKK), the police obtained some eighty armored vehicles (about 20 Ehrhardt 17/19, about 30 Daimler DZR and about 30 armored trucks).

In order to create the lacking vehicles to reach the full strength of 150 armored cars and replace vehicles no longer fit for service, the states of the country established a commission (Special Vehicle Commission), whose counsel resulted in the design and, as of 1923, the construction of so-called special vehicles, which were developed especially for police use. Three types, which looked very similar externally, were built. The Ehrhardt Automobile Works in Zella-Mehlis built 32 cars of the Ehrhardt 21 type from 1923 to 1925, the Daimler Works in Berlin-Marienfelde built 33 of the DZVR type by 1928, and the Benz Works in Gaggenau built 24 of the Benz 21 type. These special (new-type) vehicles replaced the makeshift-armored trucks and, along with the Daimler DZR and Ehrhardt 17/19, made up the armored vehicles of the police. The police of the states of the German Reich saw the armored vehicles as a source of political strength for the police, with imposing appearance and strength in action, which could be moved quickly anywhere in the state where they seemed to be needed. The special vehicles' greatest tactical disadvantages were their great weight and lack of off-road capability, which made their use on anything but main roads all but impossible.

The main reasons which could lead to the use of an armored vehicle or a platoon (two or three of them) were:

Intervention: Nipping the development of an uprising in the bud.
Scouting: Determining the situation in an area of disturbance, as a basis for police action, when other means no longer promised success because of the determination of the lawbreakers.
House, Street and Field Fighting: To relieve already fighting police, alone or together with police units of various size.
Protection of Resting or Marching Police, including those on the march in vehicles.
Sealing or Occupying tactically important points or closing off fronts during an operation.

An Ehrhardt 17/19 special vehicle in action in the central German revolt in Eisleben in March-April 1921. The first twelve vehicles of this type were contracted for in 1917 and used by the Imperial Army in World War I. Twenty more of them were built after the war.

At the beginning of the 1930s, the Special Vehicle Commission recommended that more modern vehicles be built for the police. Baden, which had not yet attained its full supply of armored vehicles apportioned by the IMKK, obtained a special Büssing NAG Type Q31P and a Mercedes-Benz Type 14/33 for testing purposes in 1931. In 1932 Prussia created a special Magirus type. All the new vehicles were more off-road capable than the old specials, but only suitable for police uses within limitations. It was mainly financial reasons that hindered the creation of new special vehicles.

Shortly after the Nazis came to power, parts of the state police force were turned into the Landespolizei, 56,000 men strong, garrisoned, and divided into Landespolizei Inspections. All the armored police vehicles were turned over to them.

But even before the Landespolizei was taken into the Wehrmacht on August 1, 1935, almost all of the special vehicles were restored to the regular police in April 1935, and not used in the Wehrmacht for training, as is sometimes stated in the postwar literature. Thus the police again had some 125 special police vehicles in 1935.

In 1936 the regular police was absorbed by the "new" Ordnungspolizei of the Third Reich, and at the same time the special vehicles were taken over by the Ordnungspolizei. The Ordnungspolizei of the Third Reich then built up its armored forces, making them into an armored troop whose police activities faded more and more into the background during the war. Aside from their security tasks, which still had a certain police character, the armored units of the police often went into battle on the front lines.

Some police armored companies, set up according to the Army's war strength instructions, differed only in the training of their personnel from similar units of the Army or the Waffen-SS.

Through the transitional army, forty Daimler towing tractors with armored bodies had been ordered in 1919 to equip the armored road-vehicle platoons. Most of the vehicles were taken over by the police as Daimler Sonderwagen DZR, like this vehicle, still fitted with iron tires, of the Hamburg security police. (ER)

In 1919, the volunteer bands of the border patrol were supplied with thirty three-ton Daimler trucks with armored bodies of chrome-nickel steel. The Prussian security police took over 14 of these armored trucks. The Berlin security police is seen practicing in 1919 with their "Moritz", which was fitted with a flame-thrower.

Shortly after World War I, such armored trucks using 4.5 ton Daimler chassis were produced for the free-corps troops. Fifteen vehicles taken over by the Prussian police were converted back into normal trucks early in 1922, since they were not suitable for police uses. (WF)

The construction of special vehicles designed exclusively for police uses began in 1923. The Daimler, Benz and Ehrhardt vehicles had almost identical armored bodies with two machine-gun turrets and a commander's turret in the middle. This Ehrhardt 21 belonged to the Saxon police. (WK)

Troop Units of the Ordnungspolizei and Their Tasks

Why did the Ordnungspolizei use armored vehicles? Why were troop units of the Ordnungspolizei established, and what were their tasks? These questions were answered very thoroughly in 1943 in the text of the Police Service Instructions (PDV) 41 "Guidelines for the Direction and Use of the Police Troops".

The PDV document will thus be quoted in excerpts or paraphrased below.

1. The armed uniformed Ordnungspolizei is a weapon-carrying representative of the state, along with the Wehrmacht and the Waffen-SS. Along with the Security Police and the Waffen-SS, it is responsible for the guarding and securing of the inner territory of the Reich, in particular for preventing and combating the unjust use of force, the protection of the state and its establishments, as well as protection and defense against attacks against the community, the people and their property.

In the war the Ordnungspolizei, in addition to their expanded tasks in the homeland area, have special tasks in the area of operations, in the backline war areas and the occupied enemy territories. Their major task herein is the pacification and securing of the lands captured by the Wehrmacht and taken into the Reich's dominions.

2. The armed uniformed Ordnungspolizei fulfills its executive tasks, as a rule, by keeping order daily. In addition it takes on tasks that can only be handled by the action of closed police bands (police troops).

The bases of the use of the Ordnungspolizei in keeping order daily are stated in PDV.27. The following statement includes the leadership bases for the use of the police troops and the anti-aircraft police with the exception of closed bands of fire police. To the extent that combat comes into consideration for the police forces as a means of carrying out their tasks, the guiding principles of the Army, as stated in HDV. 300, can be partially adopted into the document at hand. Variations from these principles are brought about by the differences existing between the Ordnungspolizei and Army in terms of tasks, structure, armament and equipment.

3. For tasks that cannot be fulfilled by individual Ordnungspolizei service, units and bands of the Ordnungspolizei formed into troops, armed and equipped, are available. Among these are the police regiments, police battalions, mounted police units, police intelligence units, police companies and platoons, motorized companies and platoons, police intelligence companies and platoons, mounted police squadrons, platoons and echelons, armored police companies and platoons, police batteries, police anti-tank companies and platoons, police grenade-launcher companies, police artillery platoons, police engineer platoons, motor vehicle echelons, repair-shop platoons, as well as special and escort commands. Their structure and equipment are ordered by the chief of the Ordnungspolizei.

4. Tasks of greater extent can necessitate the combination of several bands and units to form large police troop bands.

Along with their special use in action, the troop units of the Ordnungspolizei also serve to educate and train the replacements of the police.

5. The units of the police belong officially to a police administration (based in the homeland), and there are subordinate to the commander of the Schutzpolizei. Gendarmerie units are subordinate to the commander of the Gendarmerie. Bands and units of the police troops can, in the case of special action under higher command (outside action), be subordinate tactically and logistically to other command positions.

6. As opposed to the Wehrmacht units, which fight as a rule in the framework of great bands in collaboration of all service arms against a similar enemy, The troop units of the Ordnungspolizei take on tasks that, in their manifold and varied nature, place high demands on the ability and adaptability of their leaders, sub-leaders and men.

The basis of the police troop is the single fighter, outstandingly trained in a soldierly manner. The fulfillment of purely military tasks is a prerequisite for every unit, every leader, sub-leader and man [p. 15] of the police troop. Special training and schooling in police work and outlook built on this foundation enable leaders, sub-leaders and men of the police troops to fulfill their numerous tasks, which in similar manner are divided among the realms of military and police combat action, special securing (object protection), great surveillance service, catastrophe prevention and air protection.

7. To the extent that the police troop units were assigned military combat tasks in the framework of the Wehrmacht in exceptional cases, their subordination to the Wehrmacht lasts according to the applicable Wehrmacht directives.

On entry into battle in the framework of the Wehrmacht, the police bands and units will be, if necessary, assigned the heavy infantry weapons they lack. Their support by other service arms (artillery, tanks, aircraft, engineers, etc.) is the task of the Wehrmacht unit to which they are subordinated.

8. In war and in times of tension, the troop bands of the Ordnungspolizei may be given many various securing tasks. They primarily serve the purpose of either preventing possible attacks on especially endangered (state, military and vital-to-life) facilities and objects, or protecting the citizenry in general.

The following tactical tasks may be assigned to the troop units:

Securing buildings, objects and grounds (object protection in an extended form),

Patrolling boundaries,

Securing constructions of particular kinds (for example, roads, fortifications, military objects, etc.),

Securing supply lines of the Wehrmacht (roadways) and railway lines on special occasions (passage of special trains),

Surveillance of material and captured goods centers, prison camps and the like,

Closing off parts of cities (such as ghettos), infected herds (when epidemics break out), mined land areas, etc.,

Besides the securing of individual objects and grounds, special conditions can make it necessary for a large band of Ordnungspolizei troops can be assigned to secure entire regions (such as occupied zones). Here securing zones and sectors are to be assigned and turned over to individual police bands. They are to secure individual objects and grounds in their area according to the grade of their importance and endangerment.

9. Areas of particular unrest, above all when combat action has taken place, are to be brought to a basic pacification. The task of pacification includes making all still present enemies who took part in the previous combat unperilous, securing enemy weapons and war materials, taking over enemy command posts, propaganda and documentary materials, and destroying the basis of enemy supplying.

In pacification, the police troops can be assigned both tasks of police combat action and other police tasks (such as searching, apprehending, transporting prisoners, etc.).

10. The use of troop units of the Ordnungspolizei in large-scale inspection service serves to fulfill peacetime police tasks, which because of their extent and their distance from individual Ordnungspolizei service cannot be fulfilled alone. The impetus to imposing large-scale inspection service consists as a rule of large state and party events of all kinds, sporting events of the largest size, parades, etc. The tasks which are hereby assigned to the police troops include not only sealing off areas and other means of keeping order, securing traffic, directing crowds of people, but also upholding order and safety in the event area in general.

In war, the tasks of large-scale inspection service also expand. Traffic control and maintaining order in the Army's arrival and operational areas, carrying out removal and evacuation in enemy-threatened zones, handling large-scale human transport (such as resettling, foreign workers, etc.), can necessitate the use of troop units of the Ordnungspolizei or sections of them.

Possible Uses of Police Tanks and Police Tank Units of the Ordnungspolizei

While before 1936 the police had planned to use their armored vehicles mainly for fighting inner unrest, the purely military duties clearly dominated the armored troops of the Ordnungspolizei.

In "Weapon Technical Guidebook for the Ordnungspolizei" it is stated in the 1941 and 1944 issues, in the short chapters on armored police vehicles or police battle tanks:

> *"Armored vehicles shall combine a quick mobility (fast vehicle with good motor), sufficient safety and cover (armor), and great firepower (heavy armament with machine guns and small-caliber tank guns). The armored vehicles introduced in the Ordnungspolizei (also called armored scout cars) serve chiefly for reconnaissance. Their armament and armor make it possible to carry out their own reconnaissance, utilize good combat opportunities, and also carry out pure combat tasks."*

A training manual just for the armored police troops is not known to have appeared. In the outline of the previously cited PDV 41 there is also a chapter on materials entitled "Armored Combat Vehicles", but this has only the note "appears later". Yet in PDV 41, something can be read in the "Combat Action of the Police Troops" section (Note: In the section "Police Action of the Police Troops the armored troops are not mentioned) about the armored police troops. Thus PDV 41 will be quoted in excerpts or paraphrased below:

a) for security in motion

In order to take possession of important points on the line of march, keep sectors open or close them, or to remove obstacles, wheeled vehicles, motorcycle riflemen and police armored units, as well as machine-gun platoons on trucks, can be sent out in advance.

Forces for an advance or rear guard are to be limited to the most necessary, in riflemen to a third to a sixth or less of the total strength of the marching column. Police tanks and machine-gun units as well as motorcycle riflemen and radio troops can be subordinated.

Armored combat vehicles intended for defense drive on the line of march either before or after the marching column, or they accompany it on side roads. Rear guards are formed from rifle units, motorcycle riflemen, heavy police weapons, police tanks and special units with means of barricading.

b) On the Offensive

The focal point of the attack is marked from the start by the combined fire of all weapons, overlapping of the intelligence links, and while being carried out, by increased firing and the use of tanks and reserves. In carrying out attacks, the cooperation of motorcycle riflemen with armored police vehicles often has a positive effect.

The armored police vehicle, because of its firepower, rate of fire, armor and reverse movement, is a strong and versatile weapon. Its main disadvantages consist of its being limited to roads and paths on account of its limited off-road capability, and the difficulty of its defending itself. When an individual vehicle is used, there must in principle be other forces there to prevent its being cut off by obstacles, barricades, destroyed bridges, etc., getting into a trap and put out of action. Besides, the armored police vehicle fighting alone without supporting forces is in a position to capture ground but not hold it. In the process, the armored police vehicle is to be used advantageously for securing. In cooperation with motorcycle riflemen, it is suitable for carrying out large-scale reconnaissance and quickly taking possession of important points. Likewise it can be used to keep narrow passages open, secure flanks and rear areas, and close off enemy flanks and backlands. When attacking in street fighting, it provides excellent fire cover for the following infantry units.

c) In confrontation Combat

Armored police vehicles are assigned to advance guards, so their surprise attacks can have great effect against an unready enemy.

d) In Attacks on Positions

The possible uses of police tanks are to be made known

Points which are later important for observing one's own position, and favorable terrain for the use of police tanks, can be decisive in the choice of breakthrough positions.

The closed use of police tanks can under certain conditions make an attack easier or even make it feasible in the first place.

The mass of heavy police weapons is to be kept close together, and the use of police tanks can be purposeful.

e) With lasting resistance

On defense, armored police vehicles will seldom be available for use against lasting resistance. If they are available, they can ease the dissolution of troops falling back at times by short advances. They are also especially suitable for fire support on defense from a line of resistance and on retreat. Their armor and ability to fire all around allow them to be the last remaining resistance.

f) In Retreat

The danger of being overtaken by pursuers is to be met. For this, quickly mobile troops are of first importance. They should be equipped with police tanks and antitank weapons.

Motorcycle riflemen and police tanks remain to the last, as long as they cannot be used against the pursuers' flanks and rear. Their mobility and speed allows them to regain the connection.

g) In Combat in and around Localities (City Fighting)

Armored police vehicles can perform good service in reconnaissance and scouting in localities. It will often be possible for them to advance right up to the area occupied by the enemy. Whether they will be able to advance into the parts of the locality occupied by the enemy (reconnaissance in force) will depend on whether or not the entry roads are barricaded. Care is to be exercised in securing retreat roads.

Securing a march through and in localities must keep in mind the limited view and the possibility of surprise attacks at closest range. Armored police vehicles or ready-to-fire machine guns on vehicles can be used advantageously for watching over a march when passing through intersections.

Sending weak lateral cover or scouting troops on foot or on vehicles into side streets is not purposeful. On the other hand, armored police vehicles can be utilized for such tasks.

Since fire support in city fighting is primarily effective at short ranges, carbines can be used along with light machine guns, and at closest ranges also machine pistols. There are often favorable occasions for the use of the machine guns of police tanks.

Further information of an instructional character on the possible uses of police tanks can be found in the "Guidelines for the Use of Armored Scout Platoons" chapter of the Southeast Police Action Staff in Section III of this book.

Riflemen of the Ordnungspolizei are supported by two K-Pzkw. L6 vehicles in street and building combat in a north Italian city in the spring of 1944. The effect of the camouflage paint on the tanks can be seen very clearly.

CHAPTER II
THE FORMATION OF THE ARMORED TROOPS OF THE ORDNUNGSPOLIZEI AND THEIR FIRST ACTIONS UP TO 1942

General Information

After Himmler's installation as Reichsführer SS and Chief of the German Police (RFSSuChDtPol.) by order of July 17, 1936, he issued two orders on June 26, 1936 (RFS-SuChdDtPol.-O/S No. 1/36 and O/S No. 2/36)in agreement with Reich Minister of the Interior Frick, on the "Installation of a Chief of the Ordnungspolizei and a Chief of the Sicherheitspolizei" and "Dividing of Duties in the Operating Area of the German Police", his authority in the Ordnungspolizei and Sicherheitspolizei.

To the Chief of the Ordnungspolizei were assigned the Administrative Police, the regular Police of the Reich and of the communities, the Gendarmerie, the Fire Police and the Technical Support Police; to the Chief of the Security Police the Political Police and the Criminal Police. With that, the armored vehicles of the regular police were automatically made part of the Ordnungspolizei.

The Old Special Police Vehicles and Their Uses

On June 27, 1936 the Reich and Prussian Minister of the Interior (RuPrMdI.) had issued an order (RdErl.)(III S I d 4 Nr. 11/36 K) to the police offices and state governments about the use of the special police vehicles. In this he indicated that the investing of particular means to modernize the special police vehicles was no longer to be considered practical. In view of this, training drives in the special vehicles were no longer to be made until further notice. The police vehicles should remain unused, though taken good care of, until the final

Along with 15 Ehrhardt 17/19 vehicles, 25 Daimler DZR (this one of the Hamburg Police) were taken over as special police vehicles by the Ordnungspolizei in 1936. All vehicles of this type were released for scrapping in 1939. (WK)

These three (new-type) special vehicles of the Benz type and the (old-type) vehicle of Daimler DZR type, of the Württemberg State Police, likewise came into the possession of the Ordnungspolizei. On the command or machine-gun turrets was the warning inscription, "Clear the street or we'll shoot." (BWS)

decision as to their further use, and be kept at their regular locations. Since at various places there was no possibility of housing these vehicles, he requested [p. 19] that they notify him by July 15, 1936 at which locations special vehicles were at hand that could not be housed, and to which locations the state police could still house vehicles. He noted in particular the large numbers of these vehicles.

About a year later, on July 7, 1937, there followed a directive from the Reichsführer SS-Chief of Police in the RMDI (O-Kdo T (2) 205 Nr. 14/37), to the state governments and city police administrations, concerning the special police vehicles. In it the Reichsführer-Chief stated that he intended to reuse some twenty of the best special police vehicles of the newer type (with three turrets). To gain an overview of the numbers of still usable special vehicles, he requested that he be notified by August 2, 1937 (for the higher official agencies, July 22, 1937), giving the year of manufacture, factory, chassis and MdI numbers, and which newer-type vehicles were still in a good enough driveable condition so that in the foreseeable future no major repairs should be needed. In this evaluation the condition of the tires was also to be considered. At the same time he was to be informed whether a possible housing place for the vehicles in question was available. After receiving this information, the Reichsführer-Chief would decide which special vehicles should stay in use and issue instructions as to the further use of the special police vehicles.

Two Daimler DZVR special vehicles are seen during a drill of the Prussian police. The vehicles have different styles of paint. The drivers and commanders wear protective leather helmets.

In a directive dated November 22, 1937 (RMBliV.S. 1830) new designations and abbreviations for police motor vehicles were established. The former designation of Polizei-Sonderwagen (PSW.) was dropped and replaced by Polizei-Sonderkraftwagen (PSkw.).

The requested reports on the condition of the newer-type special police vehicles seem to have been so positive that it was decided to keep not 20 but 40 of them in service. This was probably done in view of the planned uses of large police groups, perhaps even in preparation for the planned occupation of Austria.

Thus the Reichsführer-Chief, in an RDMI dated January 20, 1938, sent a quick note (O-Kdo. T (2) 205 Nr. 2/38) to all police headquarters at which driveable special police vehicles were located. In it he stated that for parade purposes those vehicles listed in the appendix were to be put in good driveable condition by March 1, 1938 and uniformly painted Color 30 (RAL 30, Police Green). The old paint was to be left on the vehicles. In view of the material shortages (lead, etc.), the batteries in particular were to be checked for necessary repairs, and if necessary, to be sent to appropriate local firms for repairing. There were no objections if usable parts from no longer usable police vehicles were taken out and used for the required servicing. But here it was to be noted that these vehicles, because of their later use as scrap iron and steel, were to be kept in condition to be towed to the freight station. In the same way, damaged tires were to be replaced if usable spare tires were not available. If necessary, exchanges within the district could be made. The precise address for available police vehicles would be made known by the Reichsführer-Chief at a specified time. If repairing vehicles should be made questionable for lack of materials, he was to be informed immediately. He was also to be notified by February 1, 1938, with name, rank and location, how many drivers trained to drive special police vehicles were available.

In the appendix there was a list of eight Benz 21 vehicles, seven Daimler DZVR and 23 Ehrhardt 21, as well as the two Büssing NAG Type Q31P and Mercedes-Benz Type 14/33 (see Appendix 1).

In an order of February 26, 1938 (O-Kdo. O (1) 6 Nr. 3/38), the Reichsführer-Chief made known the purpose and the destination of the special police vehicles. The vehicles listed in the appendix to the order of January 20, 1938 were to be used for the Spring Parade of the German Police, and thus be delivered immediately for shipping to the Police President's Command of the Schutzpolizei in Berlin, to be picked up at the Kolonnenstrasse freight station. All necessary equipment, including attachments for machine guns, was to be included. A list of the individual parts included was to be submitted under the heading of the vehicle to which the parts belonged.

Care was to be devoted to orderly shipping that would rule out damage to the vehicle and loss of individual parts during transport.

The title "Spring Parade of the German Police" seems to be a code name for an early stage of "transport practice", the occupation of Austria, for a spring parade never took place.

Except for five Ehrhardt vehicles of the police departments in Hamburg, Halle, Magdeburg, Erfurt and Oberhausen, which were probably not sent to Berlin because of material shortages, all the requested police vehicles reached Berlin ready for use. The Berlin Schutzpolizei thus possessed 35 usable special police vehicles as of early March 1938.

Whether the police authorities, who took part in the occupation of Austria from March 11, 1938 on, under the code name of "Transport Practice", also took special police vehicles along cannot be proved for lack of sufficient documentation, but is not very likely. The police marched into Austria with four motorized marching groups, while a fifth, smaller marching group was formed for use by the Chief of the Ordnungspolizei. The special police vehicles were surely not suited to a lengthy motorized march and would have had to be shipped by rail. Whether this was planned is not definite, yet at least the fifth group of Marching Group III took along special police vehicle crews as a special formation, as shown by their diary. Marching Group III, though, was sent back to its home base in Germany between March 17 and 21.

At the beginning of the 1930s, tests were made with new special vehicle types. As of 1932 an eight-wheeled Magirus vehicle was in the hands of the Berlin Police. The modern styling shows clearly in comparison with the ten year older Ehrhardt special vehicle. (KM)

The Baden State Police in Karlsruhe tested two new Büssing (left) and Daimler (right) special vehicles as of 1931. Both vehicles were taken on by the police as individual specimens. The Büssing was produced for the Wehrmacht in somewhat changed form.

About two months later, the Reichsführer-Chief informed the applicable offices, in a memo of May 14, 1938 (O-Kdo. O (3) 1 Nr. 29/38) that the Führer and Chancellor had ordered a motorized police band formed for special tasks of the Ordnungspolizei. The police band was to be formed within the framework of the existing strength and organization of the regular police. Nothing about the normal police activities of the regular police was to be changed. The combining of staffs and units was to be prepared for and listed at once. It should be shown that within the shortest time from then, the police band or parts of it should stand ready for orders by radio.

The founding of five police brigades was foreseen, with Ordnungspolizei inspectors planned as brigade commanders. Every brigade was composed of a brigade staff and motorcycle group, intelligence platoon, music corps and ambulance platoon, and had two or three police regiments subordinated to it. Each regiment had a regimental staff with a motorcycle group and machine-gun hundred, and had three police battalions subordinated to it. The battalions in turn were divided into staff with cycle group and intelligence platoon and three hundreds.

Every machine-gun hundred (of which eleven were supposed to be formed) of the police regiments had as its third platoon a special vehicle platoon with two special vehicles (the machine-gun hundreds in Breslau, Leipzig, Hamburg and Hannover had three vehicles each), and had a strength of one officer and 26 men. Besides the two or three special vehicles, they had one motorcycle with sidecar, three motorcycles without sidecars and one Mlkw. 23 personnel carrier on hand (see Appendix 2). These third platoons of the machine-gun hundreds could also be combined as a special vehicle hundred of the brigade staff. For the time being, though, the third platoons of the machine-gun hundreds were not to be established.

On May 24, 1938 there again appeared a memo from the Reichsführer-Chief (O-Kdo. T (2) 205 Nr. 21/38) about the special police vehicles. According to it, the vehicles that were prepared according to the memo of January 20, 1938, and turned over to the Berlin police were to remain in Berlin for the time being, where they would be used for training purposes and constantly be kept in usable condition. Those vehicles maintained by the police in Hamburg, Halle, Magdeburg, Erfurt and Oberhausen but not delivered to Berlin, as well as three usable vehicles of the Koenigsberg and Tilsit police (MdI.No. 2871, 2915 and 2995) were to remain at their locations, and were likewise to be used for training and always kept in usable condition. All other special police vehicles were finally to be taken out of use and no longer maintained. All the parts of these vehicles that could be used to repair other vehicles at their own locations, such as fuel and oil lines, lights, generators, etc., could be removed. The same applied for parts that light be needed for other special vehicles still in use. Here, though, it was to be noted that the vehicles were to be dismantled only to the point where they could still roll and be towed to a freight station for shipping out later. The special vehicles taken out of service were to be housed in the available vehicle garages. Where this was not possible, they were to be placed in a corner of the premises and be protected with boards and tar paper from the worst weather and from unjust removal of individual parts. The resulting costs, which were to be kept as low as possible, were to be met from operating costs. Usable spare parts of the special vehicles were to be kept for the time being.

In an urgent message dated July 15, 1938, the Reichsführer-Chief's next message on the special police vehicles (O-Kdo.T.(2) 205 Nr. 23/38) came from the Ministry of the Interior. In view of the fact that the riding hall on Bluecher Street in Berlin, where some of the special police vehicles gathered in Berlin were stored, had to be evacuated by August 1, 1938, the Reichsführer requested, referring to the memo of May 14, 1938 (Establishment of Motorized Police Bands), that 23 special vehicles be shipped out immediately, three each to the state police in Leipzig, Breslau, Hamburg, Hannover and Stuttgart and two each to the police in Recklinghausen, Essen, Cologne and Frankfurt (see Appendix 1). All equipment was to be sent with each vehicle. The freight costs were to be covered by the receivers. The state police office in Berlin was to be assigned the Ehrhardt with registration number 2772 and the Daimler 2916 and 2993, the Benz 2999, 3004 and 3010. These vehicles were to be transferred from their previous owners to the Berlin police. In addition, Ehrhardt 2206, 2210 and 2322, Benz 2990, and Büssing NAG Type Q 31 P and Mercedes-Benz Type 14/31 remained with the Berlin police for the time being, in reference to the memo of May 24, 1938. Care was to be given to an orderly storage (see Appendix 1).

The July 15, 1938 letter includes one error. It was ordered that 23 special vehicles be ready for shipping at once, but only 22 vehicles were listed. The complete list also includes only 34 vehicles, although 35 reconditioned special vehicles were gathered in Berlin early in March 1938. The missing one is the Daimler vehicle from Potsdam, with Ministry number 2997; its whereabouts are unknown. But since, according to the May 14 order, the Hamburg police were assigned three special vehicles to form a motorized police band, it is possible that the missing vehicle was included there. But the mystery cannot be solved.

On January 30, 1939 the Reichsführer-Chief returned to the subject of the special police vehicles in RMdI. (O-Kdo.T (2) 205 Nr. 26/38 (39?). In reference to the chief;s order in RMdI.v. 24.4.38 (O-Kdo. T (2) 205 Nr. 21/38)(RMBliV.S.928 b) he reported that it was agreed with the Reich Commissioner for Evaluating Old Material, in Berlin, that the special vehicles taken out of service, according to an enclosed list (see Appendix 1) now should be turned over at no cost to the SA. for scrapping. The vehicles in question were thus to be taken, on orders from the applicable SA office in the city, to the scrapping location at no cost. Before they were turned over to the SA, their fuel tanks, fuel and oil lines and generators were to be removed if they were still present. A special directive would be sent about the later use of these parts. The turned-in vehicles were to be listed as removed from inventories. The completion of this action was to be reported to the Reichsführer-Chief.

Thus at the beginning of 1939 the Ordnungspolizei still possessed 32 (or 33) usable special police vehicles (see Appendix 1) of the Büssing-NAG (1), Mercedes-Benz (1), Benz 21 (8), Daimler DZVR (9 or 10) and Ehrhardt 21 (23) types. These vehicles were intended primarily for the establishment of special police vehicle platoons for the machine-gun hundreds. Whether these platoons were actually set up for action in 1938-39 cannot be documented. But the vehicles were certainly used for training purposes. During the Polish campaign, from September 1, 1939 on, two Daimler police vehicles saw action. The two Daimler vehicles, with the names "Memel" and "Saar", took part in war service with the State Police in Danzig (see there). Nothing is known of the later existence of the two Daimler vehicles. On November 29, 1939 the Reichsführer-Chief, in RMdI. (O-Kdo. T (G1)335 Nr. 59/39), in reference to an order of March 28, 1939 (O-Kdo. T (2) 205 Nr. 9/39), referred again to the scrapping of the vehicles taken out of service, and ordered that the equipment made available through the scrapping of the vehicles (aiming mechanisms, components of machine guns and machine pistols, bullet and water containers, ammunition drums and belts, and other components and equipment) now be delivered to the Technical Police School and Armory in Berlin SW 29, on Golssener Street.

A last written reference to the further existence of special police vehicles in the Ordnungspolizei is found in "Vehicle Technical Appendix" No. 5 of the "Special Directions for the Maintenance" No. 21 of September 30, 1942. Here, in reference to "Spare parts for special police vehicles (O-Kdo. 1 K (3) 211 Nr. 279/42), it was stated that the removal of special police vehicles was always done through individual orders. The removal had to be handled separately because the required spare parts for subsequently necessary repairs were no longer available. In the most recent times, numerous strong towing tractors were needed, and it was thus intended to reuse the chassis of the eliminated special police vehicles as tractors. To keep them usable for a long time, all spare parts still on hand should be gathered and stored at the Police School for Motor Vehicles in Vienna. Since the required storage space was not yet available, it was to be reported to the Reichsführer via the IdO. (BdO.), by January 11, 1942, where spare parts for the vehicles were still on hand and for what type of vehicle they were made. As to the shipping of the spare parts, a special order would be sent. At the time there was an urgent need for individual gears for the gearbox of a Daimler vehicle. Should gears of this type be available anywhere, they were to be reported to the Reichsführer immediately by teletype, stating their type.

As for the evaluation of material from scrapped motor vehicles (O-Kdo. I K (3) 211 Nr. 280.42), it was stated in "Vehicle Technical Appendix No. 5" that in orders to scrap vehicles of ordinary types and also special police vehicles, it was usually ordered that materials that could not be seen as replacement parts, such as copper tubes, fuel tanks, etc., were to be stored. To the extent that such materials were still at individual posts and were suitable for reuse, they were not to be removed within the framework of scrap-gathering action, but to be sent at once to the Police School for Technology and Traffic, Dept. K/D, Berlin-Pankow, Granitz Street 57-61.

Whether police vehicles were rebuilt into towing tractors unfortunately cannot be proved by either written or photographic sources. Since the chassis originally came from the manufacturer's towing-tractor production, they were certainly suited for this purpose. But since the special police vehicles were taken out of service because of a lack of spare parts, the same lack may have existed for rebuilt towing tractors.

At least a few special vehicles were still in police service until the war ended. It is shown by photos that in May 1945 a Daimler DZVR was still being used by the strengthened armored vehicle platoon in Berlin (see there).

Right: The special vehicles (new type) no longer worth repairing were, like the old-type special vehicles still on hand, turned over for scrapping in 1939. This Benz Type 21 vehicle of the Düsseldorf Police is seen being dismantled for scrapping on March 24, 1939. (STD)

Below: Most of the Ordnungspolizei's armored vehicles consisted at the beginning of the war of new-type special vehicles built by Benz, Daimler and Ehrhardt. On the "German Police Day", January 16-17, 1939, this Daimler DZVR was filmed in Grundwald. The same vehicle, with license number Pol-2673, was in the yard of the Reich Chancellery when the war ended in 1945 (see page 153). (BAK)

Steyr and Skoda Armored Vehicles and the Occupation of the Sudetenland

General Information

Since the early 1930s the German police had tried to find successor models to the Benz, Daimler and Ehrhardt special police vehicles. The experimental single examples built by the Büssing-NAG, Mercedes-Benz and Magirus firms offered no satisfactory solutions, since they were built for military use and not specifically for police use.

Nor was an armored patrol vehicle, built by Krupp and introduced at the 1938 Berlin Auto Show, accepted. This resembled a scout car built on Krupp chassis in 1933-34 by the Wilton-Fijenoord firm of Schiedam, The Netherlands.

The Ordnungspolizei almost automatically acquired modern armed vehicles in large numbers through the political and war events. After the occupation of Austria, the Austrian Federal Police, including the Security Police and Gendarmerie, became part of the German Police, being sworn in on March 16, 1938. The Federal Police in Vienna (Security Guard) at that point owned six Steyr ADGZ armored cars and three out-of-action Skoda Type PA II. In the Austrian Federal Gendarmerie, eight Steyr ADGZ vehicles were on hand. The Skoda PA II had been bought from Czechoslovakia in 1927, the Steyrs had been acquired between 1935 and 1937.

All seventeen armored vehicles were taken over by the Ordnungspolizei and gathered at the Schutzpolizei headquarters in Vienna. For the planned occupation of the Sudetenland, the eight Steyr vehicles of the former Federal Gendarmerie were modified appropriately in September 1938, automatic weapons were installed and the necessary ammunition acquired. The three Skoda armored cars, already taken out of service by the Austrian Police, were also given a general overhauling in a Vienna auto repair shop in November 1938, so as to be ready for action in the surprise march into the Sudenland.

On April 5, 1938, General of the Police Daluege and other officials from the Ordnungspolizei and Security Police commands inspected the alarm section of the Vienna Schutzpolizei at the Moroccan Barracks. The commander of a Skoda armored car (left) explains his vehicle. A Steyr armored car is at right. (BAW)

The 8th Armored Vehicle Hundred

Some of the ex-Austrian armored cars saw action with the Commander of the Ordnungspolizei (BdO.) of Northern Bohemia in the occupation of the Sudeten German area in September 1938. In Police Battalion IV/2 there was an 8th Armored Car Hundred, about the structure and equipment of which nothing is known.

The 16th Armored Car Hundred

After the occupation of the Sudetenland, the new border with Czechoslovakia was determined. Thus several new areas were occupied by Germany, and some areas were evacuated again. The occupation of a new area west of Taus took place on November 24, 1938 by the Querner Motorized Police Action Group. The Querner Action Group was composed of the Sacks Motorized Police Battalion, The I./Police Regiment 1, and the 16th (Pz.W.) Hundred. For the advance, the 16th (Pz.W.) Hundred was halved and assigned to the two battalions. This Hundred was also equipped with ex-Austrian armored cars.

Tatra Armored Cars and Rheir Action in the Protectorate of Bohemia and Moravia

General Information

Through the occupation of the Sudeten German area in the autumn of 1938 and of Czechoslovakia in March 1939, the Ordnungspolizei came into the possession of ten Czech Tatra T72 armored cars. These vehicles were previously in use by the Czech Gendarmerie and were taken into the Ordnungspolizei along with it.

For the occupation of the Czech area in March 1939, two regimental staffs had been formed to lead the involved police battalions. They were designated Police Regiment Bohemia and Police Regiment Moravia. During 1939 these regiments were renamed Police Regiment I Prague and Police Regiment II Brno.

The new Austrian armored cars first saw action with the Ordnungspolizei in the occupation of the Sudetenland in September 1938. This Steyr armored car has been marked with the police emblem, the designation WIEN (Vienna Police?), and the name "Siegfried". (ER)

From the former Czech Gendarmerie the Ordnungspolizei took over ten Tatra armored cars and used these vehicles to form armored car platoons of the Brno and Prague police regiments serving in Bohemia and Moravia. Here two of the small vehicles are seen during a break in training. (JW)

In the autumn of 1939, these three Tatra armored cars of the Brno Police Regiment were photographed. The vehicles were painted police green or Wehrmacht gray and bore the police emblem on their turrets. The crew of a small armored vehicle consisted of three men. (JW)

The Armored Vehicle Platoon of the Prague Police Regiment

According to a message from the Chief of the Ordnungspolizei (O-Kdo. O (1) Nr. 301/309) to the Reichsführer-Chief in the Ministry of the Interior early in November 1939, there were six Tatra armored cars in the Prag Police Regiment in October 1939.

On December 21, 1941 the Prague BdO. reported in a teletype (No. 351) to the deputy BdO. in Veldes – Action Staff – the arrival of an armored car platoon for southern Carinthia. The march command consisted of a commander and 13 NCO and men. The commander has been assigned three Tatra armored cars, one Mlkw. 23 Daimler-Benz, and one Skoda Pkw. 4.

The armored car platoon on the march was the former armored car platoon of the "Bohemia Police Regiment of Prague" and served to establish an armored scout platoon in the Police Action Staff Southeast in Krainburg (see there).

The Armored Car Platoon of the Brno Police Regiment

According to a report from the Chief of the Ordnungspolizei (O-Kdo. O (1) Nr. 301/39) to the Reichsführer-Chief in the Ministry of the Interior early in November 1939, there were four Tatra armored cars with the Brno Police Regiment in October 1939.

Nothing is known of subsequent use of the Tatra armored cars before the beginning of the Russian campaign.

Armored Police Vehicles in the Polish Campaign

General Information

At the beginning of the war with Poland, police battalions were put together for action as police groups (police regiments), and some were subordinated to certain army high commands (AOK.), For example, Police Regiments 1 and 2 saw action with the Eberhardt Group (AOK 3) set up in Danzig, and Police Regiment 3 (Pol.Grp. 1) was in the section of AOK 14. These police units were also assigned to armored car platoons and companies, but on account of meager sources, quickly changing organization of police forces, and the shortness of the Polish campaign, no precise information can be found.

The Armored Car Platoon of the Danzig State Police

Even before the war broke out on September 1, 1939, the secret move of two of the formerly Austrian Steyr armored cars and two Daimler special police vehicles to Danzig seems to have succeeded. The two Steyr armored cars named "Ostmark" and "Sudetenland" and the two Daimler vehicles named "Memel" and "Saar" saw action with the Danzig State Police and the SS Home Guard in their attack on the Danzig post office on September 1, 1939, being the first Ordnungspolizei armored vehicles to see war service.

Poland was aware of the danger of a Putsch by the SA, SS and Danzig Police, and had decided to offer opposition in such an eventuality. At the Polish post office on Hevelius Square in Danzig, weapons (three light machine guns, 40 pistols, and egg hand grenades) had therefore been stored secretly, so they could defend themselves for several hours against attackers. It was assumed that within a few hours the Polish Army would be able to relieve them. The German plans included the occupation of the post office on Hevelius Square by the 2nd Police District, which was housed in a side wing of the building. At about the same time as firing on the Westerplatte by the ship *Schleswig-Holstein* began at 4.45 A.M. on September 1, 1939, the Polish post office was attacked by the police of the Second District, a Hundred of the SS Home Guard, and SA men (about 150 men in all). The 56 defenders of the post fought off all attacks until late that morning. Only by using the two Steyr armored cars were the police and the SS Home Guard, supported by forces of the "Eberhardt" Group, able to make progress against the post office, put infantry guns in position, and finally, after gasoline had been pumped into the building and set afire, to conquer the post office.

One of the two Steyr armored cars was also the first vehicle to arrive in Gotenhafen. By the end of October 1939, these armored cars were being used for street patrols in the area around Danzig and to disarm Polish troops. Nothing is known of the later presence of the Daimlers , and the two Steyr vehicles were probably turned over to the Police Armored Company.

The Steyr armored cars "Sudetenland" (above) and "Ostmark" (below) were used against the Polish post office in Danzig on September 1, 1939. The armored cars in Danzig bore both the police emblem on their turrets and the skull emblem and runes of the SS. (BAK)

Ein Panzerwagen der SS-Heimwehr

When the war broke out on September 1, 1939, the Danzig State Police also had two Daimler armored cars, named "Saar" and "Memel". These vehicles also bore the death's-head as a special emblem. (PMM)

This photo is said to have been taken during the advance of the Hamburg Police Battalion 101 into Poland. The battalion was motorized in part with, and taken into action in, buses (left) of the Hamburg Traffic Agency. The crews of the Steyr armored cars still wore Austrian helmets made for tank crews, with neck protection added, and bearing the police emblem. (ER)

These Steyr armored cars, marked with German crosses but no police emblems, saw service in Poland. The crews are adventurously dressed in coats and flat beret-like caps. A small open Stkw. 2 also belonged to the tank platoon. The cover picture was based on the picture above. (BAK and RM)

The Police Armored Company of Police Regiment 3 (AOK 14)

As part of Police Regiment 3, an armored company was present during the advance of the 14th Army from Beuthen via Krakow to Przemysel. At the beginning of the Polish campaign the company, as the 13th Armored Hundred, was subordinate to Police Battalion III/3, which had been established in Vienna. The structure of the company is not known, but it was probably divided into three platoons with four Steyr armored cars each.

On September 3, 1939 a platoon for police tasks in the Krakow area was already ordered, since the city was about to be occupied. The entire company provided scouting and reconnaissance tasks in areas the Wehrmacht could not patrol, and was subordinated to various police battalions for security tasks. As of September 20, 1939, an armored platoon for security tasks was assigned to the commander of the Ordnungspolizei in Lodz.

After the renaming of Police Battalion III/3 as Police Battalion 171 on September 26, 1939, the armored company became the 5th Police Hundred of Battalion 171.

Armored Police Vehicles in the Government-General of Poland

General Information

Right after the Polish campaign, the active police forces were used in the newly formed Government-General and were also strengthened. The armored company was divided by platoons and added to the police battalions and regiments of the Government-General. What with the many reorganizations and disbanding or exchanging of the police units there, it is difficult to trace the small armored units. For the command of the police battalions included in the Government-General, the formation of four regiment staffs had been ordered on November 4, 1939. According to their locations, they were designated the Police Regiments of Warsaw, Krakow, Radom and Lublin.

An overview of the armored forces in the Government-General at the beginning of November 1939 id found in a letter from the Chief of the Ordnungspolizei (O-Kdo. O (1) 1 Nr. 301/39) to the Reichsführer-Chief in the Ministry of the Interior. According to it, there were six Steyr armored cars with the police commander in Lodz (now Krakow), four Steyrs with the Warsaw Police Regiment, and four Steyrs with the Krakow Police Regiment. These three armored car platoons were assigned to various battalions, but were overseen by a company staff. This company command was part of the Warsaw Police Regiment or one of its battalions until about mid-1940, and thereafter with the Radom Police Regiment or one of its battalions. During their service with the Government-General, the crews of the armored cars were exchanged and new crews trained in service. On March 21, 1941 the chief of the Ordnungspolizei gave out a reorganization order (Kdo.I O (3) 2 Nr. 61/41) for the Reich's Schutzpolizei in the Government-General. According to this order, the four police regiments (Warsaw, Krakow, Radom, Lublin) and still three armored police car platoons with the strength of one officer and 35 constables each were planned. The reorganization order does not indicate to which police regiment these armored car platoons were to be assigned, nor is any joint company leadership to be seen. In April 1941 there was one armored car platoon each in Warsaw and Radom; the third platoon was probably stationed at Lublin. From accounts of former members of the armored units in the Government-General it appears that the units were in service there until August 1942 and were then used for the formation of police armored companies in Vienna. The following uses and subordinations of armored vehicles in the Government-General can be documented:

Armored Police Cars with the Commander of the Ordnungspolizei

According to a letter from the Chief of the Ordnungspolizei (O-Kdo. O (1) 1 Nr. 301/39) to the Reichsführer-Chief in the Ministry of the Interior early in November 1939, there were at that time, with Order No. 15 of September 20, 1939 of the Commander of the Ordnungspolizei six Steyr armored cars assigned to the Warsaw Police Regiment.

The Commander of the Ordnungspolizei in the Government-General was stationed in Lodz until November 6, 1939, and then moved his office to Krakow. It must be assumed that the armored cars were assigned or subordinated by the Commander of the Ordnungspolizei to various police units for action.

Armored Police Cars with the Warsaw Police Regiment
(Armored Car Platoon of Police Battalion 301)
According to a letter from the Chief of the Ordnungspolizei (O-Kdo. O (1) 1 Nr. 301/39) to the Reichsführer-Chief in the Ministry of the Interior early in November 1939, there were at that time, according to a command of October 23, 1939 (O-Kdo. O (1) 1 Nr. 263/39), four Steyr armored cars assigned to the Warsaw Police Regiment. This armored car platoon was already in Warsaw in October 1939.

In the periodical "Die deutsche Polizei" No. 23 of December 15, 1939 there appeared an article about the Warsaw Police Regiment. It is stated therein: *"Along with large armored police scout cars, several modern*

The crew of a Steyr armored car has dismounted for observation. The massive towing chain draped around the cooling flap can be seen clearly. (BAK)

Three Steyr armored cars are parked beside a road. In front in one with a turret headlight and "slim" exhaust pipe, followed by another car with the "slim" exhaust pipe. The last car no longer has an exhaust pipe, but the attachment for the "thick" pipe can be seen. (BAK)

tracked vehicles of the police (armored cars) were gathered, and with them a heavy company was formed. Volunteers from the individual battalions formed the cadre of the company. When enough motorcycle riflemen are added successfully, this will be an extremely valuable means of maintaining peace and order."

The reference to modern tracked vehicles (armored cars) may refer to captured Polish 7 TP tanks, of which at least three were known to be in the hands of the Ordnungspolizei later (see Armored Platoon, Police Regiment Center).

In a late 1939 suggestion to award the War Service Cross Second Class with Swords to a Hauptwachtmeister (armored car commander) of the Armored Company, the Warsaw Police Regiment is named as the location of the Armored Company. The Reichsführer-Chief, during his visit to Warsaw on April 28-29, 1940, visited the barracks of the Armored company and Police Battalion 72 and took part in the service of those units and parts of Police Battalions 6 and 8. The battalions assigned to the regiments of the Government-General were exchanged. In November 1940, Police Battalions Warsaw I (Btl. 301), Warsaw II (Btl. 304) and Warsaw III (Btl. 307) were subordinate to the Warsaw Police Regiment. In addition to the three regular companies, the Warsaw I Battalion (Btl. 301) included a heavy police company , which consisted of a company troop, two heavy machine-gun platoons, one pursuit platoon, and an armored car platoon of one officer and 35 sergeants (SB).

Finally, an armored car platoon can be documented as part of the Warsaw Police Regiment in April 1941.

Police Armored Cars with the Krakow Police Regiment

According to a letter from the Chief of the Ordnungspolizei (O-Kdo. O (1) 1 Nr. 301/390 to the Reichsführer-Chief early in November 1939, four Steyr armored cars had been assigned to the Krakow Police Regiment at that time by Order No. 1 of September 3, 1939.

In a parade in Krakow before Reichsminister Frank on April 20, 1940, this armored car platoon appeared with its four vehicles. In the correspondence of the Krakow Police Regiment in 1940, and also in suggestions for awarding the War Service Cross Second Class without swords to members of the Armored Company in December 1940, an armored car company of the Krakow Police Regiment is mentioned.

It might be that here, as in the Warsaw Police Regiment, there was a heavy company with armored cars, or that the Armored company was subordinated to the Krakow Police Department at that time.

Police Armored Cars with the Radom Police Regiment

(Armored Car Platoons in Police Battalions 51, 111, 305 and 309)

At the beginning of June 1940, one armored car platoon was assigned to Police Battalion 51 in Pionki near Radom and one to Police Battalion 111 in Kielce. The two police battalions were subordinate to the Radom Police Regiment, which also included the company command of the Armored Company at that time. In December 1940 and January 1941, the two armored car platoons were in service at the same places, with only the battalions being switched. The armored car platoon in Kielce was now assigned to Police Battalion 305, and the platoon in Pionki to Police Battalion 309.

In January 1941, personnel of the two platoons were exchanged. The armored car drivers and several trainers of the old team stayed to train the new crews. New drivers were not trained. Steyr armored cars 7, 13, 17 and 23 can be identified as being with the platoon in Pionki. The platoon was located in Pionki to guard and secure a powder factory.

In April 1941 an armored car platoon in the Radom Police Regiment can still be documented.

Police Armored Cars with the Lublin Police Regiment

(Armored car platoons with Police Battalions 73 and 104)

To Police Battalion 104, subordinated to the Lublin Police Regiment, an armored car platoon was subordinated at least from March to June 1940. In September 1940 an armored car platoon with Police Battalion 73 in Lublin can still be documented.

Steyr armored cars are being refueled in the Kielce area. Here too, members of the crews can be seen in various uniforms. fatigue suits, cloth and leather coats and various headgear, such as Austrian helmets, steel helmets, berets and field caps can be noted. (JW)

The Steyr armored car with chassis number 26 was used by the Government-General in 1941. At least some of the crew are the same men seen in the photo at the bottom of page 30. It may also be the same armored car. (BAK)

Armored Police Vehicles in the Reich

Special (Armored) Police Vehicles
Vienna Department

At the beginning of 1939, the 17 armored cars formerly belonging to the Austrian Police and Gendarmarie were gathered into a unit by the Vienna Schutzpolizei. This special police (armored) vehicle unit, with one captain, ten lieutenants and sixty patrolmen, plus eleven special vehicles, took part in a police parade on January 29, 1939, the Day of the German Police.

In February 1939 the Vienna Schutzpolizei possessed:

1. 14 armored cars, type ADGZ, each equipped with one tank gun, one 07/12 heavy machine gun, two M. 30 light machine guns and two M. 34 machine pistols. Each vehicle was issued five M 12 Steyr army pistols and one tracer pistol.
2. Three Skoda armored cars of the older type, each with four M/ 07/12 heavy machine guns, and two M. 34 machine pistols. Each vehicle was issued four Steyr M. 12 army pistols.

Acceptance and Use of Further Steyr Armored Cars

During 1939 the Ordnungspolizei of the Wehrmacht received additional Steyr armored cars formerly of the Austrian Army. The latter had, as of 1935, obtained twelve Steyr armored cars for its armored car battalion.

According to a letter from the Chief of the Ordnungspolizei (O-Kdo. O (1) Nr. 301/39) to the Reichsführer-Chief in the Ministry of the Interior early in November 1939, the Ordnungspolizei had at this time 37 armored cars. Fourteen of them were Steyrs in use by the Government-General , and ten Tatras were in use by the police regiments in Bohemia and Moravia. There were also three Skoda armored cars (though only for the city) and two Steyrs ready for action in Vienna.

On the Day of the German Police, January 29, 1939, a parade was held in Vienna, and vehicles of the Special Police (Armored) Vehicle Unit took part. This photo shows the Steyr special vehicles being prepared for the parade on January 28. (BAK)

Eight Steyr armored cars were not ready for action at that time: One was at the Technical Police School in Berlin for testing purposes, one was at the Steyr factory, and six were undergoing repairs.

On November 10, 1939 the Chief of the Ordnungspolizei notified the command office that the Reichsführer SS had ordered that the two Steyr armored cars that were ready for action in Vienna be sent to Berlin immediately. At the time, no armored cars were needed in Vienna, and in any case the three Skodas would suffice. Of the six cars being repaired, two were to be stationed in Berlin and the other four in the Ruhr area. There is no further correspondence about this on hand. Nor is it clear where the last two Steyr armored cars of the former Austrian Army were at the time, for there were 26 armored cars in Austria in all. After November 1939 these two vehicles were also taken over by the Ordnungspolizei, for in a comparison made by the Chief of the Ordnungspolizei on the state of equipment on August 20, 1940 with the beginning of the war, it is stated:

Armored cars of the modern type:
9/1/39: 37 Pzkw. – 8/1/40: 30 Pzkw.

Besides the 39 modern armored cars, there were naturally older special police vehicles still in use by the Ordnungspolizei. To avoid confusing these armored cars, the Reichsführer-Chief at the Ministry of the Interior wrote particularly of the designation of police vehicles in a message (O-Kdo. K (1) 201 Nr. 40/40) on July 1, 1940:

"(1) In the message of 11/22/1937 (RMBliV.S. 1830) the designation and its abbreviation for police vehicles is determined. The designation included there, "Pol.-Sonderkraftwagen" (PSkw.) applies from now on only to the motor vehicles of this type already present in the old Reich before 1938. For the added Steyr, Skoda and Tatra vehicles, the designation of Panzerkraftwagen (Pzkw.) applies effective immediately.
(2) In all reports and communications the aforementioned abbreviations are to be used. The difference between PSkw, and Pzkw. is strictly to be applied."

The Austrian Army had in its armored car battalion twelve Steyr armored cars in service, designated M 35 medium armored cars. These vehicles were taken over by the Wehrmacht and only turned over to the police in 1939. This photo of seven Steyr armored cars and Wehrmacht soldiers was taken in 1938. (KM)

Armored Police Vehicles in The Netherlands

General information

Nothing is known of the use of armored police units during the western campaign.

In The Netherlands the Ordnungspolizei, after the establishment of a BdO for the country at The Hague in June 1940, had to be assigned armored scout cars captured in the western campaign. Exactly when the Ordnungspolizei received these vehicles, how many and what types they were, and which unit was first supplied with them, cannot be determined.

In any case, the Ordnungspolizei received vehicles of the "Pantserwagen L 180" type, an armored scout car built by the Swedish firm of Landsverk, of which The Netherlands had obtained twelve in 1936. Photos show two of these "Holland armored cars" of the Ordnungspolizei in a parade in The Hague on February 10, 1941.

In addition, one "Wilton-Fijenoord" armored scout car was obtained by the Ordnungspolizei. The Wilton-Fijenoord firm of The Netherlands had built this vehicle on a Krupp L2H43 chassis in 1933-34 for use in Java. Three vehicles were built, two being sold to Brazil in 1935. The third one had been shipped to Java in 1934, but was not usable there because of its too-heavy weight, and had been returned to The Netherlands. There it is said to have been at an artillery arsenal near Hembrug (Zaandam) in May 1940, without armaments. From there it went into the possession of the Ordnungspolizei.

Armored Car Platoons in Police Battalion 41

Police Battalion 41 was assigned to The Netherlands by order of the Reichsführer-Chief on June 3, 1940 (O-Kdo. O (1) 1 Nr. 415/40). By his order of July 22, 1940 (O-Kdo. (3) 2 Nr. 205/40), the battalion was assigned two armored car platoons. Nothing is known of the platoons' strength or equipment. In November 1940 Police Battalion 41 was stationed in Amsterdam.

Armored Car Platoons in Police Battalion 254

By order of the Reichsführer-Chief on November 25, 1940 (O-Kdo. O (1) 1 Nr. 797/40), the replacement of Police Battalion 41 by Police Battalion 254 was ordered as of December 16, 1940. Police Battalion 254, with home barracks in Koenigsberg, was prepared for this service by order of the Reichsführer-Chief on October 29, 1940 (O-Kdo. O (3) 1 Nr. 124/40). The driver staff and the two armored car platoons of Police Battalion 41 remained with Police Battalion 254. The armored car platoons stationed in Amsterdam became part of the 4th Company of Police Battalion 254.

Armored Car Platoon(s) in Police Battalion 68

In May 1942, Police Battalion 68 conducted a training course for crews of armored cars from The Netherlands. Twelve patrolmen (SB) of the motor vehicle service were ordered to take this course. A training certificate has been found, certifying the completion of the training as a driver for an armored vehicle designated "Netherlands Krupp Armored Car". This indicates that the captured "Wilton-Fijenoord" armored scout car at that time was assigned to Police Battalion 68 in The Hague. Since fifteen men were certainly not trained as drivers for a single armored car, there must have been other armored vehicles from Holland on hand with Police Battalion 68.

On February 10, 1941 these two Pantserwagen M-36, formerly of The Netherlands, took part in an Ordnungspolizei parade in Holland. The vehicles had been captured in May 1940. They are marked with the police emblem and the numbers 1 (car in front) and 2 (car in back).

Armored Police Vehicles in the Russian Campaign

General Information

In the Army Groups North, Center and South established for the Russian campaign, the Commander's Staffs 100, 101 and 102, later renamed North, Center and South, were formed to patrol the military back land areas.

For each of the military backline areas a "Higher SS and Police Leader" (HSSPF) staff was established for police tasks, and a motorized police regiment was subordinated to it. The police regiments were made up of a staff with two armored car and two antitank gun platoons, one intelligence and one technical emergency service company. At first, three police battalions were subordinated to them at first. The subordinated units varied.

As the front moved eastward, the occupied areas moved out of the Backline Army Area Commanders' responsibility to a civilian administration. Commissariats-General were created. The HSSPF were responsible for the safety and order of these areas.

The HSSPF for Russia-North (General of the Police Jeckeln), with its seat in Riga, was simultaneously assigned to the Reich Commissioner Eastlands as the HSSPF Ostland. The BdO of Riga was subordinate to him. To the four commissariats-general the SSPF and KdO were assigned, for Estonia in Reval, for Latvia in Riga, for Lithuania in Kauen and for White Ruthenia in Minsk (belonging to Russia-Center as of 4/1/1943). In larger cities there were SS and Police commanders (SSPStOF.), and in the country there were SS and Police District Leaders (SSP-GebF.) in charge.

The HSSPF for Russia-Center (General of the Police v.d. Bach-Zelewski), with its seat in Mogilev, was renamed HSSPF Russia-Center and White Ruthenia as of 4/1/1943. Subordinate to it were the SSPF of Mogilev and of Minsk, plus four SSPSTOF in Baranovici, Smolensk, Mogilev and Vitebsk.

The HSSPF for Russia-South (General of the police Prützmann), with its seat in Rovno (in Kiev as of February 1942), was also responsible for the Reich Commissariat of the Ukraine. For the territory of Army Group A, an HSSPF Black Sea was established in November 1943. This was subordinate to the HSSPF Russia-South, which was renamed HSSPF Russia-South and Highest SSPF Ukraine in March 1944. In the Territory of the Reich Commissariat Ukraine there were ten SSPF and KdO in service: SSPF Volhynia (Brest-Litovsk, as of September 1942 Lusk), SSPF Shitomir, SSPF Kiev, SSPF Nikolayev, SSPF Dnepropetrovsk, SSPF Chernigov, SSPF Kharkov, SSPF Tairia (Simferopol), SSPF Stalino, SSPF Rostov, SSPF Avdeyevka, and SSPF Pripyet (Pinsk).

Armored Car Platoons in the Police Regiment North

In the schematic war structure of the Chief of the Ordnungspolizei in June 1941 concerning the police units subordinate to the HSSPF North, Center and South, two armored car platoons were assigned to each HSSPF. The schematic structure shows an armament of three over heavy machine guns (or tank guns), 15 heavy machine guns and six machine guns for every platoon of the Center and South regiments. No armaments are listed for the platoons of the North Regiment.

The schematic war structure for the Commander, Backline Army Area 101 (North) of July 22, 1941 includes three armored cars, each armed with six heavy and six standard machine guns, for each of the two armored car platoons of Police Regiment North.

The schematic war structure for the commander, Backline Army Area North, of December 19, 1941, shows three armored cars, each armed with four machine guns, for each of the two police armored car platoons of Police Regiment North.

From this information there is no possible indication of the structure and the armored car types of the armored car platoons.

But photos are available, showing one platoon of three Steyr armored cars, and one platoon of three Tatra armored cars in action with Police Regiment North. They were accompanied by sidecar motorcycles with machine guns. If every armored car had a motorcycle with a machine gun, then each platoon was armed with three tank guns, 15 heavy machine guns installed in armored cars, and six machine guns on sidecars. It may be possible that double or false entries were made erroneously in the schematic war structures, but this is not definite. Only sporadic actions or action areas of the armored car platoons of Police Regiment North are documented. Until the end of 1941, Police Regiment North was subordinate to the Commander, Backline Army Area North.

Besides the Steyr armored cars, there was also an armored car platoon of three Tatra armored cars, shown here in the center of the photo. The Steyr armored car in the foreground clearly shows a swastika painted on the turret top for identification from the air. (BP)

Two Steyr armored cars of police Regiment North are accompanied by motorcycle riflemen. On the sidecars of the first two cycles there is painted a springing dog (or big cat), apparently the emblem of Police Regiment North. (CWB)

An armored car crew observes action in the field from the edge of a forest. Note the German crosses painted above and below on this Police Regiment North vehicle, plus the white markings on the fenders.

As of October 7, 1941, the regimental staff, with the intelligence unit, armored car platoons, antitank platoons, and the 3rd Company of Police Battalion 321, moved from Pleskau to Opachka. As of January 10, 1942, the police Regiment North was in action on the front, as the "Keuper" Group, subordinated to General Command of the Tenth Army Corps, and fought along with the 81st Infantry Division of the Army until March 1942. The "Keuper" Group, with its staff and a scout car platoon, saw action, for example, in Kakilevo from January 19 to 22, 1942. The other units served in Rachutcha, Yevaknovo, Nagovo and Bubkovskshina.

From April to June 1942 the "Kepuer" Group was subordinate to the 18th Infantry Division (mot.) within the Tenth Army Corps and defended a sector called Parfinostrasse in the Staraya-Russa area. On April 24, 1942 the 18th Infantry Division (mot.) reported, in a radio message to the Tenth Army Corps, about the heavy weapons of the armored car unit of the "Keuper" Group. Listed were five light machine guns; two heavy machine guns and two 3.7 cm antitank guns were no longer mentioned and were perhaps no longer usable. On June 18, 1942 the Commander of Police Regiment North wrote in his situation report to the HSSPF North:

"Since all attempts to refresh the regiment have failed, I saw myself compelled to make the following report to the 18th I.D. (mot.) on June 18: According to the command from the RFSS sent to me by the HSSPF, the police troops made available to the Wehrmacht are to take on only such tasks as they are able to achieve on the basis of their training and constitution.

The command requires that the leader of the formation make an appropriate report at the right time.

In reference to my report of June 14, 1942, re the fighting value of the Police Regiment North, I must therefore report, as is my duty, that the Police Regiment North, what with its present strength, which cannot be completed, and reduced combat value, resulting from ceaseless action without any possible relief, is no longer suited for front duty in its present form."

In an inventory, filled out on July 8, 1942, about the Infantry Assault Medal in silver, the troop unit equipped with armored vehicles was described, as in a radio message of the 18th I.D. (mot.) on April 24, as "Pzkw.-Pak.-Abtl." The armored car and antitank platoons thus seem to have been combined into one unit.

In July 1942 the staff of Police Regiment North was used to form Police Regiment 15.

No written evidence remains as to the fate of the armored cars, but photos document that remaining vehicles were sent back to Vienna. The armored car crews of the former Police Regiment North returned to the police administration of Vienna or to the Police Vehicle School by November 1942.

Armored Car Platoons in Police Regiment Center

The schematic war structure of the chief of the Ordnungspolizei for June 1941 for the police units subordinated to the HSSPF North, Center and South shows that each HSSPF was assigned two armored car platoons. The schematic structure showed every platoon of the regiment to have an armament of three over heavy machine guns (or tank guns), 15 heavy and six standard machine guns.

The same armament was shown in a schematic war structure of the Police Regiment Center, as described by the Commander of Backline Army Area 102 (Center) of June 24, 1941.

Photos taken in July 1941 show the existence of at least one platoon with three Steyr armored cars.

This armored car platoon was stationed in Warsaw in July 1941. The three Steyr armored cars bore the names "Memelland", "Danzig" and "Pommern" on their sides. In addition, they were adorned with oversize swastikas.

The 221st Security Division, subordinate to the Commander of the Backline Army Area, recorded in its war diary on July 8, 1941:

"Through the Police Regiment Center, three armored scout cars are made available on request. The armored scout cars are assigned by the division to Security Regiment 2 for use as pursuit commands on runways 1 and 2."

The armored car platoon with its three Steyr armored cars arrived in Bialystok in August 1941 and in Slusk in September 1941.

The existence of a police armored car unit in the Police Regiment Center area as of July 1941 is also known from photographic material. It was equipped with one platoon of armored cars from Holland (two vehicles) and one platoon of armored cars from Poland (three vehicles).

These units, designated Armored Scout Car Platoon and Armored Battle Tank Platoon, were assigned, with an antitank platoon, to the 10th (heavy) Company of Police Regiment Center. When this heavy company was established is not known. These armored vehicles were also marked with large swastikas.

Actions and exact action areas of the armored car platoons of Police Regiment Center can be documented only sporadically. When the Russian campaign began, the Police Regiment Center was subordinate to the Commander of Backline Army Area Center and sometimes assigned to the 221st Security Division. On August 22, 1941 Police Regiment Center left the command area of the 221st Security Division, and served from August 23 to September 1, 1941 under the Commander of Backline Army Area 580, seeing service against partisans.

In an indefinitely defined operation in the Backline Army Area Center, the 10th (heavy) Company, beginning on September 9, 1941, used its armored scout car platoon to secure the Mogilev-Bobruisk road from the Dubrovno fork to the heights of Borki. In July and August 1942 the Police Regiment Center was subordinate to the commander of Smolensk. A schematic war structure of the Smolensk Command shows for the 10th (heavy) Company a structure of one platoon with two armored cars (h), one platoon with three tracked armored vehicles (p), one platoon with three antitank guns (r).

Steyr armored cars were not mentioned, and their presence is indefinite; the vehicles were probably returned to Vienna. The staff of Police Regiment Center was used to form Police Regiment 13 in July-August 1942. Probably the 10th (heavy) Company also became part of Police Regiment 13 and renamed Armored Company "Center" (see there).

Vehicles of an armored car platoon of the Police Regiment Center are seen at the end of August 1941 in Bialystok (above) and Slusk (below). Along with the police emblem on the turrets, they have large swastikas painted on the turret tops and the sides of the vehicles. The armored cars also bear names like "Danzig" and "Memelland". (PH)

Three 7 TP armored cars of the Police Regiment Center in the autumn of 1941. Along with the German crosses front and rear on the upper body and the police emblem on the turret sides, a big swastika was painted on the bow plates of these tanks. (BAK)

A formerly Polish 7 TP armored car of the Ordnungspolizei, with a messenger's motorcycle at its right side. Note the two vivid white crosses on the rear. To the left of the armored car is a film car of the Berlin Technical Police School. (BAK)

Besides the Polish 7 TP armored car with 3.7 cm tank gun, the unit also had at least two armored cars from Holland, bearing the names "Den Haag" and "Amsterdam". (BAK)

The armored cars from Holland were painted with unusually large police emblems but bore no German crosses. The spotlight on top of the turret was added by the Ordnungspolizei.

Armored Cars of Police Regiment 2 Russia-Center (later Police Regiment 14)

(Heavy Company, Police Regiment 2 Russia-Center, later 13th (heavy) Company, Police Regiment 14)

The leader of the police forces (essentially Reserve Police Battalions 51 and 122) in Operation "Potsdam" in the Kirov-Kostnichki-Olsa-Beresina-Dimanovchshina-Bobroisk Road-Mogilev area on June 10-11, 1942 already had a heavy Russian armored scout car on hand and wrote about it in his combat report:

"The heavy weapons (tanks) created in the framework of the police forces proved themselves splendidly and were able to relieve the riflemen of their difficult reconnaissance task. With the expansion of these means of combat I promise myself good success in the fulfillment of combat tasks."

A short time later, on July 7, 1942, the chief of the Ordnungspolizei ordered the establishment of a second police regiment for Russia-Center, formed of:

Reserve Police Battalion 51 as First Battalion

Reserve Police Battalion 122 as Second Battalion

Ukrainian Constabulary Battalion as Third Battalion and a heavy company.

The structure and equipment of the heavy company of Police Regiment 2 Russia-Center is not known. From regimental commands and action reports of July 16-17, 1942, though, it appears that at least the following weapons were on hand:

3 tracked armored cars

2 tracked towing tractors

4 armored scout cars (heavy and light types)

1 light infantry gun battery (2 guns, horse-drawn)

2 antitank guns.

The heavy company was housed at the Militz Barracks in Mogilev. The heavy company, with Police Regiment 2 Russia-Center, saw service in the Mogilev area, among others, to secure the Mogilev-Dovsk road.

By order of the Reichsführer-Chief on July 9, 1942 (O-Kdo. I O (3) 1 Nr. 184/42), Police Regiment 14 was established. The regimental staff, Reserve Police Battalions 51 and 122, and the Heavy Company came from the renamed Police Regiment 2 Russia-Center. Along with the restructuring and renaming of Police Regiment 2 Russia-Center, the renaming of the Heavy Company as the 13th (heavy) Company, Police Regiment 14, also took place, by order of the Regimental Commander on August 8, 1942.

According to a report of September 8, 1942, the 13th (heavy) company was equipped with:

2 armored cars

4 heavy armored scout cars

1 light armored scout car

4 light infantry guns

2 antitank guns

The company had a strength of two officers and about 90 men, with only the cadre personnel coming from Germany, the rest being Ukrainian policemen. The latter were also used, for example, as drivers. The 13th (heavy) Company with its armored vehicles was now a part of Police Regiment 14:

- From July 19 to August 2, 1942, in Operation "Adler" (partisan action in the Belinichi, Kolichenka, Gorodiche, Sagatye, Dolchenka, Podkoselye, Veliki-Log, Belyi-Log, Olen area).
- From August 14 to 26, in Operation "Greif" (partisan action in the Orsha, Krasnoye, Lyesno, Vitebsk, Smolyany area).
- From September 2 to 7, 1942, in Operation "Breslaü (partisan action in the Smoliza-Dobusha area), losing one heavy armored scout car to a mine, one heavy armored scout car to infantry fire at the engine, one light armored scout car to engine damage.
- from September 23, 1942 on, partisan action at the former troop training camp northwest of Kolichenko.
- From October 4, 1942 on, in Operation "Regatta".
- From November 20 to 26, 1942, in Operation "Nürnberg".

As of November 27, 1942, Police Regiment 14 was transferred to the front in the area of the Italian Trentino Division. On December 15, 1942 the regiment had to transfer to Ivanovka at the small Don bend, where it was wiped out within days in withdrawal fighting.

The 13th (heavy) Company had stayed in Mogilev except for its antitank platoon, and did not see front action. On order of the HSSPF Russia-Center, it and other remnants of Police Regiment 14 were combined into a company.

Nothing is known of the fate of the armored cars and scout cars, but it is possible that they were used to equip the 12th Police Armored Company.

Police Regiment 14 was officially disbanded in March 1943 and established anew (see also 2nd Police Armored Company).

A heavy Russian BA 10 armored scout car is shown in front, with a light Russian BA 20 scout car to the rear. Emil was the tank commander's first name. (EG)

Parts of the 13th (heavy) Company of Police Regiment 14 are loaded on a train. Two T 60 armored cars, two Russian (BA 10) heavy scout cars, and one light Russian (BA 20) scout car can be recognized. (EG)

Armored Car Platoons of the Police Regiment South

In the schematic war structure of the Chief of the Ordnungspolizei re the police forces subordinate to the HSSPF North, Central and South in June 1941, two armored car platoons are assigned to each HSSPF. The schematic structure indicates for every platoon of the Center and South Regiments an armament of three over heavy machine guns (or tank guns), 15 heavy and six standard machine guns. The existence of at least four Steyr armored cars in the Police Regiment South can be proved by photographs. Little else is known to date of the Police Regiment South and its armored car platoons. A radio message from the HSSPF Russia-South to the RFSS command staff on October 5, 1941 stated: "Police Regiment South strengthened by armored cars and antitank guns and used to fight scattered Russian troop units."

The areas where the armored car platoons of the Police Regiment South saw action were Brody, Busk and Peresyaslev.

In July 1942 the staff of Police Regiment South was used to establish Police Regiment 10.

The armored cars still on hand were probably returned to Vienna.

Armored Scout Cars of the Wehrmacht for the HSSPF North, Center and South

On August 18, 1942 the Reichsführer-Chief sent a message to the Chief of the Ordnungspolizei:

"Through the intervention of the Reich Security Headquarters, SS Group Leader Streckenbach, we have to date received 13 armored scout cars from the Wehrmacht for the purpose of fighting bandits. I am having three of these armored scout cars sent to Minsk, five to SS Obergruppenführer von dem Bach, and five to SS Obergruppenführer Prützmann. They are being assigned and delivered to the Ordnungspolizei. SS-Ogrf. von dem Bach and SS Obgf. Prützmann, though, are receiving from me personally the assignment to turn over at least two of these armored scout cars in the next months, in order to gather practical experience, to the Sipo for their investigative operations in surveillance of the bandits. A report is to be made to me by December 31, 1942 on the use of the armored scout cars in the service of the Sipo."

These thirteen armored scout cars may have been captured Russian vehicles. The type of scout cars and their locations cannot be verified. Possibly this number also includes the cars that, for example, were used by Police Regiment 2 Russia-Center, established in July, (later Police Regiment 14), and the Armored Company Center.

Motor vehicles of the Police Regiment South are seen on the market place in Busk. Along with many cars and trucks, four Steyr armored cars are also parked here. Two Steyrs are ready for use, with open hatches and doors; the other two are covered with tarpaulins. (EH)

An armored car platoon of Police Regiment South pauses during a march. Since the vehicles' weapons are fitted with dust covers, they do not seem to be going into action. (EH)

In September 1942 this Steyr armored car of the new series was photographed in the Shitomir area. At that time no armored police company was in action in that area. It could be that the first new-series Steyr armored cars were made available to the HSSPF North, Center and South for troop testing. (DIZ)

CHAPTER III
THE EXPANSION OF THE ARMORED TROOPS OF THE ORDNUNGSPOLIZEI SINCE THE END OF 1941

General Information

Besides the "normal" police tasks, the main task of the HSSPF was securing the occupied territories, especially those in Russia, by fighting against partisan bands. Here the HSSPF led its own actions (operations) through, or worked with, the Army's security units or other troops. In crises on the front, individual police troops were used for "stabilizing", or formed into whole battle groups of division strength and used as such.

For fighting partisans and action at the front, better and better-equipped large motorized units were needed, which finally resulted in the formation of police regiments. A command of the Reichsführer-Chief on July 9, 1942 ordered the merging of all police and reserve police battalions into regiments. There were 28 regiments planned, and the staffs needed for them were formed gradually.

Each regiment was supposed to include the following regimental units: intelligence company, engineer platoon, armored car company and antitank company, later also a police gun company. These regimental units, probably for lack of equipment, were never established in sufficient numbers. Most regiments were never assigned an armored car company, though others had two. A few armored car companies were led as independent units and never became part of a regiment. The same was true of some antitank and gun companies.

The HSSPF Russia-Center and White Ruthenia was given full authority by the RFSS, on October 3, 1942, to fight against partisans, and on June 21, 1943 he was named Chief of Partisan Fighting (Pol.). One reason for this was the fact that partisan activity in Russia took place more and more in the boundary areas between the HSSPF Russia-North (Eastland)/HSSPF Russia-Center (White Ruthenia) and Russia-Center (White Ruthenia)/Russia-South (Ukraine), which made central planning and command necessary.

Armored Scout Platoon of Police Action Staff-Southeast

Establishment, Structure and Equipment

On December 21, 1941 the BdO. of Prague reported in a message (No. 351) to the executive BdO. in Veldes-Action Staff that an armored car platoon had been sent on the march to southern Carinthia. He referred to a message from the Reichsführer-Chief in the Ministry of the Interior (No. 1973) of December 20, 1941 (O-Kdo. Ia K I Nr. 136/41 g). The marching command consisted of a leader and three sub-leaders and men. Three Tatra armored cars, one Daimler-Benz Mlkw. 23, and one Skoda Pkw. 4 had been assigned to the command. Transport by rail to Graz was to take place in the night of December 21-22; from there they would march by road via Maribor to Krainburg, where they were to report to the Police E-Staff Southeast. The armored scout platoon sent on this march was the former armored scout platoon of the "Böhmen" Police Regiment.

In an order of December 29, 1941 (O-Kdo. I K (2) 209 Nr. 117.41), the Reichsführer-Chief arranged for motorcycle groups from the Dresden IdO Zone to march with the three armored cars on their way to the Pol.-E. Staff Southeast. There were six motorcycles and nine more with sidecars (perhaps BMW), with drivers and eighteen police officers (SB) to be sent on the march to Krainburg at once, under their own power. The policemen were to be armed with carbines and equipped with three field glasses, three flare pistols, three machine pistols, three machine guns, three pistols, 500 hand grenades, three despatch pouches and 108 flashlights with 108 batteries. The motorcycle groups were sent on the march on January 1, 1942.

A Tatra armored car is looked at by officers and men of the Ordnungspolizei in front of the post office in Krainburg (?). The small vehicles had three-man crews and were armed with two machine guns. (BAK)

The commander of a Tatra armored car of Scouting Troop I shows two officers how the turret machine gun works. Near the left elbow of the officer at the right, the emblem can be seen, and the I in a circle on the front fender is the sign of Scouting Troop I. (BAK)

After the motorcycle rifle groups arrived, the Pol.-E. Staff Southeast set up an armored scout platoon. The armored scout platoon was to be ready for action by January 14, 1942. The armored scout platoon was part of the motor vehicle echelon of the Pol.-E. Staff Southwest and was structured as follows, according to an order from that staff of January 21, 1942:

Leader Group
Armored Scout Troop I   (1 Tatra armored car)
Armored Scout Troop III   (1 Tatra armored car)
Armored Scout Troop V   (1 Tatra armored car)
Armored Scout Car VII (reserve)   (1 Tatra armored car)

The leader's group consisted of the platoon leader, his deputy, and two drivers, who had a car, a personnel truck and a motorcycle available.

Every armored scout troop was composed of a motorcycle rifle group and an armored scout car. The motorcycle rifle group consisted of the group leader, who was likewise the leader of the scout troop, the light machine gunners 1 and 2, gunners 3 through 5 with their carbines, and five motorcycle drivers. There were three motorcycles with and two without sidecars. The Tatra armored scout car had a commander, who was also the deputy scout troop leader, a gunner and a driver.

The fourth armored scout car and its crew formed a reserve and took the place of any other armored scout car that was put out of action, or could be used to strengthen an armored scout troop. Five other policemen were divided into reserve armored car crews and replacements for lost motorcycle drivers or gunners.

Why four and not three armored scout cars were on hand is not explained in the available sources, but the existence of all four cars is proved by photos. The numbering of the scout troops with Roman numerals I, III, V and VII is also unclear.

When the 14th Armored Police company was established, Armored Scout Troop I became the First Armored Car Platoon in this company (see there). Just when the 14th Armored Police company was established is not known. The first order to establish it is dated July 5, 1943, and the process was finished by the end of October 1943.

Action

The first quarters of the armored scout platoon were at Villa Sitters in Veldes, but the platoon was transferred to Krainburg as of January 21, 1942. The platoon was supplied by the staff company in Veldes and was subordinated to its company chief in terms of discipline. The platoon was officially supplied and paid by Reserve Police Battalion 181 in Krainburg.

On May 25, 1942, new permanent quarters were assigned and the armored scout platoon was divided: The Leader's Group and Troop I were located in Veldes, Troop III in Krainburg, and Troop VII in Goreinavas. The permanent quarters were changed again on August 1, 1942: The Leader's Group and Troops I and III were stationed in Veldes-Seebach (Koenigsgaragen) and Troop VII in Pölland. The location of Troop V is not stated. The area of action for the armored scout platoon was Carinthia, but no records of individual actions exist. The police Action Staff Southeast. though, set up guidelines for the use of the armored scout platoon on January 3, 1942 (revised on January 21, 1942), which are given here in revised form, as they give an excellent portrayal of the action of an armored police scout platoon.

Guidelines for the Action of the Armored Scout Platoon

1. After the arrival of 3 motorcycle rifle groups, an armored scout platoon (Pz.-Spähzug) will be established in the Police E-Staff Southeast. The Police Scout Platoon will consist of 3 armored scout troops.

2. The armored scout troop consists of one "Taträ armored scout car with 3-man crew and a motorcycle rifle group (Kschtz.-Gr.). One Kschtz.-Gr. will be composed of the scout troop leader (his deputy is the commander of the armored scout car), 5 riflemen and 5 motorcycle drivers.

3. The armored scout troops of the armored scout platoon are always tactically subordinated to the units.

4. The armored scout troop is to be used as a unit for reconnaissance and informational tasks. For patrolling and securing tasks the armored scout cár and motorcycle riflemen can be used separately.

An armored scout troop in training during the winter of 1942-43. In front is the Tatra armored car, followed by the riflemen on their solo and sidecar motorcycles. (TF)

The Tatra armored car of Scout Troop V, with its emblem on the right front, in snowy country. Tools and equipment are carried on top of the rear fender. (TF)

5. The armored scout car is to attract the enemy's notice and defenses to itself when the armored scout troop has been noticed. For that reason it takes the leading position in missions of the entire armored scout troop as a matter of principle. On routes in plain sight, the motorcycle of the scout troop leader (leader of the motorcycle rifle group) follows in second position, at a distance of 100 meters. The other cycles follow between 75 and 100 meters back. On routes not in plain sight, the cycles follow within sight of each other. Here the speed of the armored scout car may not exceed 30 kph.

6. The motorcycle rifle group may receive special tasks, such as:

   a. Removal of obstacles (barriers, etc.)

   b. Securing the carrying of messages to the assigning units.

7. All obstacles are to be viewed at first as occupied by the enemy and are thus to be overcome.

8. Inclusively applied and under the personal command of the motorcycle rifle group leader, the rifle group shall come at the enemy's flank and rear, while the armored scout car is to occupy the enemy frontally with fire from cover and provide cover for the attacking rifle group. Effect is more important than concealment. Broad deployment, best utilization of terrain, ruthless use of weapons with all means upon confrontation with the enemy, and bold attacking assure success to the dismounted rifle group. With good view and possibility of effect on the enemy it will often be necessary to carry out the following: The armored scout troop drives far back into cover. From there the motorcycle rifle group is applied as in number 7, while the armored scout car engages the enemy with fire in a second advance and provides fire cover for the attacking rifle group.

9. If it becomes known that the resistance cannot be broken with one's own means, the armored scout troop must break away from the enemy. The decision to do so is up to the armored scout troop leader, who leads the dismounted motorcycle rifle group. For communication of this decision, he can use light signals, the meaning of which is to be determined before each undertaking. Likewise the meeting place after breaking off from the enemy is to be agreed on as the effective point of departure.

10. The drivers remain with the motorcycles to secure them and cover the riflemen's retreat. The motorcycles are handled as in #12.

11. In tough fights and difficult positions, the motorcycle messenger may have to bring reports to the rear without the usual riflemen and without orders.

12. Obstacles not occupied by the enemy are to be disposed of by removal. At impassable bridges, fords are to be located and marked. Securing is to be kept in mind here, and carried out by the armored scout car and light machine gun. The removal personnel do not put down their weapons. During this activity the motorcycles often stand pointed in the opposite direction with motors running.

13. Driving through localities with unknown situations is to be done under fire protection by the armored scout car and the motorcycle rifle group. It will often be practical to have the rifle group drive around the localities.

14. The task of the armored scout troop is reconnaissance and scouting, not combat. It is only to be used in removing obstacles, otherwise avoided.

15. Structure of the armored scout troop:

An armored scout troop consists of an armored scout car and a motorcycle rifle group,

   a) the motorcycle rifle group consisting of a group leader, likewise leader of the armored scout troop, the light machine gunners 1 and 2 and the gunners 3-5 as carbine carriers and close-range fighters, and five drivers.

   b) a "Taträ armored scout car with commander, likewise deputy armored scout troop leader, a driver and a gunner.

16. Crews of the motorcycles: [/ divides two lines in book]

   Sidecar Cycle No. 1: Driver / Rifleman 3 Group Leader

   Sidecar Cycle No. 2: Driver / Gunner 2, Gunner 1 with light machine gun

   Sidecar Cycle No. 3: Driver / Gunner 4, engineer tools and ammunition (according to setup)

   Solo Cycle No. 4: Driver / Gunner 5

   Solo Cycle No. 5: Driver (Messenger)

17. Equipment:

   a) The Motorcycle Rifle Group

Group Leader: Machine pistol, flare pistol, telescope, message pouch, flashlight
Gunner 1: Light machine gun, pistol, light machine gun ammunition
Gunner 2: Carbine, spare barrel, light machine gun ammunition
Gunner 3: Carbine
Gunner 4: Carbine
Gunner 5: Carbine
Every Driver: Carbine
Every man of the rifle group: / 2 hand grenades and sidearm
Pistols are carried if they are available.
b) The Armored Scout Car crew
Commander: Pistol and light machine gun
Driver: Pistol and carbine
Gunner: Pistol and light machine gun
Every armored car also carries a flare pistol with five white, five red and five green flare cartridges, and 25 hand grenades.

18. *Dress*

   a) The motorcycle rifle group: Service dress (jacket, breeches, boots), belted raincoat, steel helmet, goggles, gauntlets. Wearing the coat under the raincoat, and wearing head protection, is always to be ordered by the leaders.

   b) The armored car crew: Tank dress, boots, belt with only pistol, crash helmet, gloves.

19. *Each motorcycle rifle group will carry these engineer tools: one saw, one hatchet, one spade, one pick, one towrope, two balled and two stretched charges, and explosive shells if ordered.*

20. *The following ammunition is carried:*

   a) Motorcycle rifle group: to be ordered

   b) Armored car: 300 rounds of SS ammunition, 25 hand grenades.

21. *The armored scout platoon is a part of the motor vehicle echelon of the Staff Company/Pol.-E.-Staff Southeast.*

   Its quarters are in Krainburg. Its members are disciplined and supplied by the Staff Company, with the exception of food and salary, which are to come via the Res.-Pol.-Batl. 181.

22. *The action of parts of the armored scout platoon will be reported via the Ia to the company leader of the staff company, who arranges anything further. Requirements of subordinated units and assignments of parts of the armored scout platoon are to be directed to Ia.*

An armored scout troop of the pol.-E.-Staff Southeast. Each of the three armored scout troops had one Tatra armored car and several motorcycles. The Tatra armored car had a three-man crew and was armed with two machine guns. (BAK)

Armored Car Unit of the Vienna Schutzpolizei

On May 27, 1942, exiled Czech agents assassinated the Chief of the Reich Security Headquarters and SS Obergruppenführer and General of the Police Reinhard Heydrich, authorized representative of the Reich Protector for Bohemia and Moravia. Heydrich succumbed to his injuries on June 4, 1942.

Because of the assassination, an order of the Vienna Police Command (Ia 5170/Pzkw.) of May 28 for May 29, 1942, called for the special use of an armored car unit. On May 29 the fourth training course for Steyr armored car crews, given by the Central Motor Vehicle Staff of the Vienna Police administration, was to begin. Because of the attack on Heydrich, though, the beginning of the course was postponed, and from the personnel of the motor vehicle staff, the trainees who arrived on May 28, and armored and other vehicles of the Central Motor Vehicle Staff, an armored car unit was assembled and sent on the march to Prague and its vicinity for outside service.

The Armored Car Unit had at least five Steyr armored cars. The undertaking, begun on May 29, 1942, was designated "Special Action Heydrich" and ended on June 12, 1942. The Armored Car Unit was then assigned to the cities and areas of Brno, Prague and Pilsen. No participation by the Armored Car Unit in reprisals for Heydrich's assassination, including the taking of hostages or the destruction of the village of Lidice on June 10, 1942, can be proved,

Whether the Armored Car Unit was assembled especially for the "Special Action Heydrich", or whether all the armored cars located in Vienna were combined into one unit with this name, cannot now be proved for lack of source materials. But in 1939 all the armored vehicles located in Vienna had been gathered in the Police Special Vehicle (Armored) Unit Vienna (see there), and an armored readiness unit with trained personnel was part of the Central Motor Vehicle Staff in 1942. After the special service ended, the fourth armored car training course was conducted by the Central Motor Vehicle Staff from June 15 to July 24, 1942.

Steyr armored cars in front of the Brno railway station early in June 1942. (AW)

Vehicles of the Armored Car Unit at a barracks in Prague in June 1942. To expand their radius of action, four spare cans of gasoline are strapped on above the right rear wheel. (AW)

Steyr armored cars in Pilsen during "Special Action Heydrich" in July 1942, with some of the crews standing in front of the vehicles for a photo. (AW)

Police Armored (Car) Unit

Establishment, Structure and Equipment

As of January 1942 the founding of a police mountain regiment was already foreseen, and at that time it was planned to assign this regiment an armored unit. In a list dated February 18, 1942 for planned housing for the Police Mountain Regiment to be organized, an armored car company with 14 armored cars, two cars, 5 personnel trucks., 20 supply trucks, 14 motorcycles and 161 men was included.

But only on May 23, 1942 did the Chief of the Ordnungspolizei give the order (Kdo. I O (3) 1 Nr. 102/42) to establish a police mountain regiment at once. The mountain regiment was at first directly subordinate to the chief of the Ordnungspolizei. Along with the regimental staff and Police Battalions 302, 304 and 325, the following regimental units were to be set up:

A heavy police mountain company, a police engineer platoon, a police cavalry platoon, a police intelligence unit, and a police armored car unit.

Garmisch-Partenkirchen became the home base of the regimental staff and most parts of the regiment. The intelligence unit was based at Greinau and the armored car unit at Mittenwald.

The Police Armored Car Unit of the Police Mountain Regiment was to consist, according to the Strength and Equipment Directive "Pol.-Pzkw.-Abt./Pol.Geb.Jäg.Rgt." of May 1, 1942 (see Appendix 4), of the following:

Armored Car unit staff including workshop platoon and supply column

1st Platoon Pzkw. (3 Steyr armored cars)

2nd. Platoon Pzkw. (3 Steyr or Netherlands (?) armored cars)

3rd Platoon Pzkw. (5 Renault armored cars)

4th Platoon Pzkw. (5 Renault armored cars)

The Police Armored Car Unit of the Police Mountain Regiment, according to this strength statement, had 16 armored cars, two cars, 6 Stkw., 23 trucks, 21 cycles and 150 men.

Why this armored unit was designated an "Abteilung" although in numbers of men and vehicles it did not differ significantly from the company originally planned, is not known.

The Reichsführer-Chief ordered in a directive of June 4, 1942 (O-Kdo. I K (2) 251 Nr. 61/42) that trained armored car crews be transferred from their bases in Germany to the Motor Vehicle and Tank School in Vienna, in order to assemble a police armored unit for service with the Police Mountain Regiment. The involved forces were to report by 8:00 PM on June 15, 1942 to the Motor Vehicle and Tank School of the Ordnungspolizei, Landstrasser Hauptstrasse 68, in Vienna.

Nine Renault and two Steyr armored cars are seen at the unit's parking lot in Krainberg. The Renaults have the police emblem painted on the right front and left rear of the turret. The size difference of the two vehicle types is clear to see. (ML)

Discussing the situation beside a Steyr radio car of the Police Armored Unit during training in the Garmisch area in the summer of 1942. Only one second-series Steyr armored car was fitted with this large antenna, which was also used on the six- and eight-wheel radio cars (Sd. Kfz. 232) of the Wehrmacht. (BAK)

While the situation discussion is held by the Steyr radio car, the platoon's two other Steyr armored cars secure the intersection at the rear. The radio car itself secures to the front, so that no unnoticed approach on the road is possible. (BAK)

Their supply of uniforms was to come from their home bases, according to Appendix 1, pertaining to service outside the country, of the order of July 12, 1941 (MBliV. S. 1303) and the additions to it added later, while their special clothing, described in Section M of Appendix 1, PBKlV. Part II, was to come from Vienna. Their home bases were to provide one 08 or 7.65 mm pistol with 80 or 50 bullets respectively, an 84/98 bayonet and a gas mask.

It was also pointed out that the organization must under all circumstances be carried out in the planned manner, because it concerned specially trained personnel, and replacement forces in this field were not yet available.

After the establishment and training of the armored car unit in Vienna, it was transferred to the Police Mountain Regiment in the Garmisch-Reutte-Ehrwald-Leermoos area and quartered in Mittenwald until mid-July 1942.

As of July 1942 the Police Mountain Regiment bore the number 18, and the armored car unit was correspondingly also designated as Police Armored Unit 18.

Action

As of July 26, 1942, Police Mountain Regiment 18 was transported to see service in the Oberkrain. The stationing information for Police Mountain Regiment 18 of August 1, 1942 names Krainburg as the post of the armored car unit, with the regimental staff also quartered there.

In the Oberkrain, the regiment carried on anti-partisan warfare in various operations.

August 3-4, 1942: at Blegos

August 7-8, 1942: in the Poklyuka near Veldes

August 10-14, 1942: on the high plateau of Yelovka (armored unit took part)

August 19-22, 1942: in the Stony Alps

August 29-30, 1942: in the Tuchein Valley

To what extent the armored unit took part in these operations is not known, but photos show that it took part in an operation on the German-Italian boundary area from August 31 to September 4, 1942.

Further operations of the Police Mountain Regiment in the Oberkrain were:

September 8-10, 1942: in the high plateau of Jelovka (armored unit took part)

September 13-14, 1942: in the Herzogswald (armored unit took part)

September 25-26, 1942: in the St. Valentin-Moräutsch area

Here too it is not known whether the Police Armored Unit was always involved.

As of October 4, 1942 the regiment was withdrawn to the Garmisch. Reutte- Ehrwald-Leermoos area. It was then transferred by rail to the Danzig area beginning on December 1, 1942, and then by ship to Finland.

The Police Armored Unit was not sent along to Finland, but separated from Police Mountain Regiment 18 on December 8, 1942 and, as of December 9, 1942, joined Police Regiment 2, formed on orders from the chief of the Ordnungspolizei of September 2, 1942 (Kdo. I O (3) I Nr. 213/42) in the areas of the HSSPF Russia-Center and White Ruthenia. The time of the Police Armored Unit's transfer had been determined earlier, for a message from the Reichsführer-Chief on December 3, 1942 ordered the assigning of another administrative official to Police Regiment 2, stationed at that time in Slonim. The regiment was made up of other regimental units, including among others an armored unit.

From January 6 to 26, 1943 the armored unit took part in partisan fighting in the Ofipovice area. An armored platoon of the unit was assigned to the SS Dirlewanger Special Battalion and was with it in Sluzk on January 19. On January 11, 1943 Police Regiment 2 reported to the Chief of the Ordnungspolizei in Berlin that Armored Unit 18 had arrived without administrative officials and requested that one be assigned.

In a message to the Reichsführer-Chief on February 18, 1943, Police Regiment 2 again requested the assigning of more administrative officials and based this on the increased manpower of the regiment. Among others, an armored unit (194 men) and a workshop platoon (13 men) had been added. The armored unit and the workshop platoon were stationed in Minsk.

Renaming

On the basis of documentation, the following designations can be proved:

    -Armored Unit, Police Mountain Regiment   -1st Police Armored Unit, Police Regiment 2

    An order for the renaming of the armored unit cannot be found to date, but among Field Post number 14322 "Pol.Pz.Kw.Abt." in the field post directive applicable as of January 25, 1943, the unit is listed as "Pol.Panz.Kp.(verst.) 1".

Above: A second-series Steyr armored car is seen on a training trip in the Garmisch area. In front, the BMW messenger's cycle belonging to the scout car can be seen. (BAK)

Left: A Steyr armored car of the Armored Car Unit on a reconnaissance mission. It is followed by two messengers on motorcycles, who can take the results of the reconnaissance to the scout platoon leader or the unit leader. Since the Steyr armored cars were not equipped with radios, messengers had to be utilized. (BAK)

Two second-series Steyr armored cars are seen parked in Krainburg. The engine cooling flaps are closed and the turret weapons have been covered. No bow machine gun was installed in these vehicles, and the opening for the gun mount was covered with an armor plate. (ML)

The five Renault tanks of an armored platoon are seen in Krainburg. Their registration numbers can be seen to the right front of the bow, as well as the unit's emblem (an edelweiss?) at the front of the right track apron. Behind the Renaults are the three Steyrs of an armored platoon of the unit. (ML)

Renault tanks take part in a practice march in the Garmisch area. The ex-French tanks were taken over by the Police Armored Car Unit without technical changes. Only the paint and emblems indicate the new ownership. (BAK)

The driver of a Renault tank demonstrates the meager entry and exit space in the small armored vehicle. He wears an unpeaked leather tank cap. The hinges of the exit hatch can be seen. Next to the hatch is a horn. (BAK)

Renault tanks on the march to prepare for partisan action in the Oberkrain. The first vehicle has set up a signal flag, which was used to transmit commands among the tanks.

Covered by a Renault tank, troops of Police Mountain Regiment 18 search a yard in the former Yugoslavia (Oberkrain). A disadvantage of the French tanks was that the commander had to look out the turret hatch to have a good view. To be able to observe with more armor protection and get to the weapons faster, the small turret cupola was later cut into and fitted with hatch covers.

1st Reinforced Police Armored Company

Establishment, Structure and Equipment

After the order for the establishment of a police mountain regiment had been issued on May 23, 1942, the Reichsführer-Chief directed, in an order (O-Kdo. I K (2) 251 Nr. 61/42) of June 4, 1942, that a police armored car unit be established for the Police Mountain Regiment.

In July or August 1942 the formation of the 2nd Police Armored Company began. It reached Russia for service as of October 1, 1942. After the 2nd, the 3rd, 4th, 5th and subsequent police armored companies were formed. A 1st Police Armored Company has not been found in the available documents before the renaming of the armored car unit at the beginning of 1943.

The 1st (reinforced) Police Armored Company formed by renaming the Police Armored Unit in January 1943 had the following structure:

Group leader including workshop platoon and supply train
1st Armored Car Platoon (3 Steyr armored cars)
2nd Armored Car Platoon (3 Steyr or Netherlands (?) armored cars)
3rd Tank Platoon (5 Renault tanks)
4th Tank Platoon (5 Renault tanks)

The exact equipment of the 2nd Platoon has not been confirmable to date from either documents or photographs.

In a letter to the Reichsführer-Chief on February 18, 1943, Police Regiment 2 requested that further administrative officials be assigned to it, basing this again on the increased manpower of the Regiment. Among others, an armored unit (194 men) and a workshop platoon (13 men) had been added. The Police Armored Unit, subordinate at that time to Police Regiment 2, had already been renamed as the 1st (reinforced) Police Armored Company.

An armored scout car of the company has gone off the road and sunk in. A rescue attempt has been attempted, with towing by the other two scout cars. The armored police companies had no special rescue or towing vehicles. The workshop platoon had one towing tractor with a low-loader trailer. (WP)

The increase in manpower to 207 including the workshop platoon, up from 150 men in the structure of May 15, 1942, could be explained by the addition of a motorcycle rifle platoon (see Appendix 8) and perhaps an intelligence platoon. No documentation of this could be found. From the end of March to mid-June 1943, armored cars of the 1st (reinforced) Police Armored Company were being repaired in Vienna. It could not be determined just how many vehicles and which types there were.

Action

The company began its service with Police Regiment 2 in the HSSPF Russia-Center and White Ruthenian area, where it fought partisans, still designated Police Armored Unit. At the beginning of 1943 it was renamed the 1st (reinforced) Police Armored Company. The following actions in the parameters of Police Regiment 2, and at times in the "von Gottberg" Battle Group, have been documented:

June 6-26, 1943: Partisan fighting in the Ofipovice area. At times one platoon was subordinated to the SS Special Battalion Dirlewanger.

January 27-March 19, 1943: Fighting large bands in the Sluzk area and the Pripyet Marshes.

March 20-21, 1943: Partisan fighting in White Ruthenia.

March 24, 1943: Special Operation "Ruda-Javonkä in the Ruda-Javonka area.

May 1-10, 1943: Operation "Draufgänger II" in the Rudnia-East Grodek-Teja-Manili Wood area. Police Regiment 2 with subordinated Schm. Btl. 118, SS Special Battalion Dirlewanger, and Police Armored Company 12.

May 12-June 28, 1943: Operation "Cottbus" in the area west of Lepal.

July 2-5, 1943: Operation "Günther" in the Logoisk-Rudnia and Manili Wood areas.

July 12-August 11, 1943: Operation "Hermann" in the Koadnov-Kamiov-Rudnia Naliboki-Perebnyno Naliboki area.

August 13-15, 1943: Operation "Skank" in the area southeast of Koidanov.

August 20-September 15, 1943: Partisan fighting in the Begoml-Vokshyie area.

November 5, 1943: Attack on Sarachie.

November 6-7, 1943: Attack and breakthrough to Dretun oil storage depot.

On January 10, 1944 the 1st AK took command of all troops and bands in the Polozk-South Jasno Lake area. Among them was the "von Gottberg" Police Battle Group. In a report from this battle group dated January 10, 1944, the company is listed as being in Police Regiment 2 as the 1st (reinforced) Police Armored Company, with a strength of six heavy eight-wheeled armored cars and nine tracked armored vehicles. One more of the latter seems to have been lost.

The same applies to a report of the "von Gottberg" Battle Group as of March 1, 1944. On March 25, 1944 the 1st AK of the 1st Police Armored Company was reported as having left the corps area.

The 1st (reinforced) Police Armored Company took part in Operation "Kormoran" and was in the Borisov area on June 26, 1944. In July 1944 Police Regiment 2 saw service in the "von Gottberg" Battle Group under the command of Gen.Kdo. VI AK.

From there the police Armored Company was summoned by verbal orders from the Ordnungspolizei Headquarters to leave combat service in the east and return to the school in Vienna for replenishment of men and supplies.

The overview of the Ordnungspolizei forces and their actions as of July 15, 1944 listed the company as being in the area of the BdO Vienna, although the Police School for Motor Vehicles did not report the company's arrival in Vienna until July 23-24. The overview of the Ordnungspolizei forces and their actions as of October 20, 1944 also names the 1st Police Armored Company as being in the BdO Vienna, HSSPF Danube area. Nothing is known of any actual added manpower or new supplies of the 1st Police Armored company, but on February 22, 1945 the company returned to outside service.

At the war's end the 1st Police Armored Company was serving in Oberkrain. The Commanding General of Backline Army Zone E reported in a message to Army Group E about the locations of the most important troop units on April 26, 1945 that SS Police Regiment 17, SS Police Regiment 28 "Todt", with the 1st and 3rd Battalions of SS Gendarm Battalion 3 (mot.) and the 1st and 14th Police Armored Companies were in service on the Goettenitz-Gottschee to Gurk (Topla Reber) line.

Protected by their Steyr armored scout car, the crew repairs a wooden bridge. The Steyr armored cars were only conditionally off-road capable and were directed to use more or less paved roads or good solid ground. (KL)

The paved road runs on a causeway, but the wooden bridge over a gap is destroyed. The Renault armored car gets around the destroyed bridge near the causeway. The driver is helped through this difficult maneuver by an observer on the bow of the vehicle. (WP)

Parts of the armored company are seen in a Russian village. At the right front is a Peugeot truck, with another in the background, followed by two motorcycles and a Renault tank. (HL)

The Renault seen in the photo above is shown close-up here. The small turret cupola was no longer in its original French form when photographed in June 1943, but already fitted with a two-part cover. (HL)

Covered by a Renault tank, police riflemen move forward against partisans. Two riflemen are seated on the rear of the small tank.

The tank seen at the bottom of page 68 is shown from the right side. It can be seen that the tank commander can observe through the new opened cupola and no longer has to expose his upper body outside the armor. (PFA)

A camouflaged position for a Renault tank secures an open area at the edge of a forest. Note the makeshift tent top that suggests the position was planned for longtime use. This Renault is already equipped with a radio set, as the antenna shows. (PFA)

A British Matilda II infantry tank that reached Russia through the lend-lease program is inspected and perhaps tested by men of the armored company. (WP)

An antitank weapon scar on the windshield of this Steyr armored car of the 1st (strengthened) Police Armored company shows how its driver lost his life in the spring of 1943. (WP)

2nd Police Armored Company

Formation, Structure and Equipment

The 2nd Police Armored Company's date and order of establishment are not known to date. It was formed, according to notes in personal papers, as of July-August 1942.

The 2nd Police Armored Company was made up, according to the strength and equipment instructions of July 25, 1942, as follows:

Leader's group including supply train and workshop platoon

1st Armored Car Platoon (2 Steyr armored cars)

2nd Armored Car Platoon (3 Steyr armored cars)

3rd Tank Platoon (5 Renault tanks)

The 2nd Armored Car Platoon of the 3rd Police Armored Company, according to an order from the Reichsführer-Chief (O-Kdo. I K (2) 251 Nr. 110II/42) of August 4, 1942, had to be turned over to the 2nd Police Armored Company after being formed. The reasons for the transfer of this platoon are not known.

The 2nd Police Armored Company was stationed in Vienna-Knödlhütte (in a small hotel near Purkersdorf) during its formation, from August 10 to September 30, 1942. The company's personnel numbered 118 men.

Action

As of October 1, 1942 the company was sent to the district of the HSSPF Russia-Center, arriving in Mogilev on October 7, 1942. It was subordinated to Police Regiment 14 and quartered at 90 Dniepr Street in Mogilev. A single action report from the 2nd Police Armored Company (dated October 8-20, 1942) is the only such report delivered, or at least discovered, to date, and is given here in its full text:

Armored Company   Mogilev, October 23, 1942

Police Regiment 14

Action Report (excerpt)

10/8/1942. Mogilev, Dniepr Street gathering place in a block of flats.

At 1:00 P.M. a discussion takes place at the SS Police Leader's office, at which the use of the company for fighting partisans is ordered. The command was not to be upheld because our arrival had not yet been assured. At first it is only clear that the company is to be directly subordinate to the leader of the opera-

On October 1, 1942 the armored cars and other motor vehicles of the 2nd Police Armored Company were loaded for transport at the Unterpurkersdorf depot near Vienna. (AW)

tion, Brigade Leader and Major General of the Police von Treuenfeld. The Batt.z.b.V./Pol. 14 sets out on the march on October 9, 1942. The armored company is to provide covering fire from Belinichi on.

10/9/42, as before.

The armored platoon makes its preparations for the departure on 10/10/1942.

10/10/42, as before.

The Armored Platoon has received its marching orders. It departs at 6:00 A.M. and picks up the Batt.z.b.V. on the Mofilev-Beresino road, some 6 km southwest of Belinichi, around 10:00 PM. Quarters for the night of October 10-11 will be in Sababini. The Company (without the 3rd Platoon) has further work service, fills up with fuel, and receives marching rations for ten days. The mood is excellent, the service is accepted joyfully.

10/11/42, Mogilev, 90 Dniepr Street and Mogilev-Beresino road.

After a talk by the chief of Staff, the Company (without the 3rd Platoon) marches out at 1:00 P.M. At this time it gets the assignment to escort a radio truck to Kulakovka. But since the place is about 8 km off the line of march, it is left at the approach to Kulakovka. The usual small technical disturbances occur (locking brakes, loose wheels, electrical problems). One armored car has radiator damage, which has occurred several times in Vienna already. After our service it will have to go to Vienna for repairs. About 6:00 P.M. the Company arrives in Novoselki and takes quarters. The platoon leader of the 3rd Platoon reports after his task is finished. He had already reached the town about 1:00 P.M. Here one gets the impression that the population is thoroughly friendly to the partisans. The Starost carries out every task successfully and quickly. He tries to do the German salute. Groups of homeless children are not present.

10/12/42, Novoselki.

Clean weapons, make repairs to vehicles and equipment. On request the Starost delivers potatoes, cabbage, and a calf for the company's food. In talks it turns out that the population has a violent hatred of the partisans, who took all their grain and bread, so that potatoes are their only food for the winter. The young people were taken away. The people's trust can be won easily through honest, straightforward treatment. The fuel supply column has arrived. At 9:00 PM the company chief received the command to report to the brigade leader. See Appendix 1 (see experience report).

10/13/42, Novoselki.

The brigade leader has arranged for the company chief and his troop to be transferred to Beresino, while the company at first remains in Novoselki. At 1:00 P.M. the move takes place, with 1/5, 1 Stkw.(4), 2 cycles and one sidecar cycle. About 2:25 P.M. the Kpzkw. Platoon is ordered to Kukarevo, about 14 km from Novoselki. As far as we know it is to advance the march of a company of the II./SS I.R. 8, which had to draw back because of enemy action during an advance in the forest. A surrounding attack is carried out. Code names are "grass gnat" for the brigade, "climbing rose" for II./I.R. 8. At 4:00 P.M. a motorcycle courier brings a report to Novoselki that a towing tractor broke down with wheel trouble after going some 10 km. At 4:07 P.M. the repair shop platoon leaves. At 5:50 P.M. a report says it is not just a wheel, but axle and bearing damage. Repairs cannot be made at this time, since no spare parts of the kind are available. At 10:00 P.M. the company chief receives the order for the next day's action, when the company will be subordinate to SS Infantry Regiment 8.

10/14/42, Novoselki.

The company has been given the task of reconnoitering the forest paths off the Beresino-Mogilev road in the direction of Kukarevo with the 1st and 2nd platoons' armored cars. In the morning the 8th Company suffered heavy losses in the region northeast of Kukarevo.

Along with that, the guard on the road is to be strengthened. The 3rd Platoon carries out the task of supporting an advancing company (6./I.R. 8). Both operations take place without combat. In the evening the company takes up quarters in Vyachenka. The night passes without notable events.

10/15/42, Vyachenka.

The 3rd Platoon moves from Ilovka to join the Company.. In examining the route to take toward Micheyevici in the afternoon, the company chief finds [p. 73] that a small bridge in Shurovka must have been broken down by the inhabitants after the 13th and Flak Infantry Regiments had marched through. After reporting

this to the regiment in Meskovici, the company stays in Shurovka, while it was originally to go to Dmitrovici.
10/16/42. Shurovka.

The regimental order for 10/16/42, included as an appendix, ordered the Company to Dmitrovici. In the discussion, cooperation of the Armored Company with the 14th Company in an attack on a forest camp southwest of Krasnoye. It does not succeed, as the camp and the still-known positions were reported as abandoned. The armored car platoon reaches the camp without incidents. After completing the march to Micheyevici, the 3rd Platoon's supply train is ordered back to Micheyevici. Over night the 1st and 2nd Platoons were used to secure the road northwest of Pogosst. Since that time they have been lost, since the company chief has received no notification of further action. An abandoned armored car that could be repaired is ordered for securing alone by the brigade leader. He reports beck to Shurovka at 11:00 A.M. on October 17.

10/17/42, Shurovka

A courier sent to the regiment by the company brings the information that the Armored Car Platoon stays in Shurovka today and presumably will begin a march of 75 km on 10/18/42. Vehicles are prepared accordingly. I. R. 8 is silent as to the action of the two armored car platoons. The Company Chief is concerned as to whether the fuel supplies will be enough after the action to date. About 2:50 P.M. the 1st and 2nd Platoons arrive back in Shurovka. During the whole night they worked in close collaboration with the Motorcycle Rifle Company of I.R. 8 to guard a sector of some 6 kilometers, ending at 7:30 A.M. on 10/17/42. At 7:50 A.M. Oblt. Sch. receives orders to do reconnaissance with a platoon on the conditions on the Kukarevo road northeast of the Hazverk road. After carrying it out without incident, an appropriate report is sent to the Chief 14.I.R. 8, At 4:50 P.M. the commander of I.R. 8 requests escort for Batt.z.b.V. Pol. 14 to Vassilievstchina. Battalion is on the march at once. 1st Platoon departs 4:55 P.M.. At 3:00 P.M. the brigade receives orders as in Appendix 2, by which the company again is subordinate to the brigade and is to report for duty in Chepelvici by 6:00 P.M. on 10/18/42. Company is given the order to march out.

About 5:30 P.M. a Ukrainian reports, stating that he belongs to the Mogilev militia barracks. Name: Vassily D. He will stay with the company until it returns.
10/18/42, Shurovka.

The march to Chepelvici brings unexpected difficulties. While going through a village, deep filth, which all the vehicles get through. At a ford, though, all the cars and trucks get stuck in a swamp. If it hadn't been for the company tractors, we would never have reached our destination, likewise other combat formations. The 75-km march has lasted eight hours. At 3:45 P.M. we arrive in Chepelvici. At 5:00 P.M. orders come to go off to Smogilovka. We arrive there about 6:30 P.M. Find quarters. Orders are to follow. Two couriers are sent to brigade. One man sick with apparent kidney infection.
10/19/42, Smogilovka.

In the night before 10/19/42, heavy fire in a westward direction is experienced several times. During the day no orders arrive from the brigade to the company. Chief 14./I.R. 8 states on his return that the brigade staff has been sent some 40 km to the north.
10/20/42, Smogilovka.

Since no orders arrive during the morning, and no combat formations can be discerned in a radius of some 20 kilometers, the Company Chief decides to seek information himself. One prisoner is to be sent to Mogilev. On the trip the Company Chief meets the Batt.z.b.V., Police regiment 14 near Belenici and learns that the action has ended. Then he returns to Smogilovka and leads the company back to its quarters in Mogilev on 10.21.42.

Thus the report of the action of the 2nd Police Armored Company with SS Infantry Regiment 8 in Operation "Karlsbad". The Company Chief was not satisfied with the tasks assigned to his company, for in the end he was simply forgotten. Many of the command difficulties can be traced to the company's meager radio equipment. The Company Chief wrote an experience report about the action in Operation "Karlsbad"; it too has survived and is given here, since it reveals the basis for the action of the police armored companies and also contains important information on the technical readiness for action of the Steyr armored cars and Renault tanks:

<u>Experience Report</u> about the action of 10/12/20, 1942, Operation "Karlsbad"

The position under the brigade has not worked out well for the company. It has only been assigned the task of reporting daily at various places. The Regiment Commander of SS I.R. 8, on two days, gave the two scout platoons the task of guarding the Mogilev-Beresino road or increasing the securing by the 14./ I.R. 8. The intentions of the higher command were not to be understood from the received orders to march. On the other hand, the Company Chief could not make any suggestions for the action, since he was constantly with his formation and did not know the sequence of combat events. Thus he got the feeling of being held back as an unnecessary reserve.

In these movements the armored scout cars traveled an average of 580 kilometers. The tanks covered 410 kilometers. With these figures one should keep in mind that the life span of the motors of these scout cars, with the best care, is no more than 10,000 km, and those of the tanks no more than 3,000 km. Thus such actions cannot be carried out more than 18 or 7 times respectively.

The tanks were called on with the basic concept that thanks to their partial or full off-road capability and their firepower, they could open the way for the advancing troops, or at least form their backbone. But in this action there was no clarity on this point. They were applied only from the periphery and to the outside.

The practical use of the company is thus only possible when the leader of a battle group constantly controls the tanks assigned to him, is at liberty to utilize them in foreseen focal points, and then directs the tank leader. To keep all parts of the company ready for use for as long as possible, the attempt must always be made to avoid long march distances.

On October 30, 1942 it was recorded in the regimental diary that the former designation of 2nd Police Armored company was dropped and the company was immediately to be designated the 14th (Armored) Company, Police Regiment 14.

On November 7 and 8, 1942, the 2nd Police Armored company or 14th (Armored) Company, Police Regiment 14, along with the regiment, took part in the storming of Stariza. Here one armored car was fired on and set afire, and there were numerous wounded, including the Company Chief.

In November 1942 the armored company again took part with Police Regiment 14 in attacks on Pelchauszk and Ozierovo, and in Operation "Nuremberg" from November 20 to 26, 1942. Operation "Nuremberg" was carried out by the "von Gottberg" Battle Group, and the 14th (Armored) Company was subordinated to the III./Police Regiment 14, for example, in the fight for a crossing on the Dzisna and the ensuing action on November 23, as well as an attack on Stovrovno. On November 27, 1942 the units of Police Regiment 14 (I./ 14, III/14, 13th (Heavy), 14th (Armored) and 15th (Intelligence)) received orders in Veydelavka to march by rail to action at the front at once. The destination was unknown at first. After seven days of unbroken rail travel, news came through Army Group B that all motorized units were to unload at Kubyansk, some 80 kilometers east of Kharkov, and all other units in Valuki, some 120 km eastnortheast of Kharkov. From these two points they were to reach Rossosh by overland march. The motorized units arrived there on December 10 and 11, 1942. Police Regiment 14 was to be subordinated to the "Trentinö Division after consultation with the Italian A.O.K. 8, and relieve Italian troops (Alpini Battalion "Edolö) on the northern Don on both sides of Bassovka. The replacement date was 6:00 A.M. on December 16, 1942, and the positions had already been viewed by advance commands. When Police Regiment 14 reached Propovka on December 15 and was still some 45 km from the positions it was to take, an alarm order reached it around noon, canceling all the plans and ordering all usable units of the regiment to march 95 kilometers overland to Ivanovka at the smaller bend in the Don.

Some units reached Ivanovka around 6:00 P.M. on December 16 (1st, 2nd, 9th, 10th, 11th (only the Heavy Machine Gun Platoon), 13th (only the Antitank Platoon), 14th (Armored) and 15th (Intelligence) Companies), and a temporary regimental command post was erected at the northern exit from Ivanovka. Briefing on the situation was provided by the Ia of the 385th Infantry Division of the Wehrmacht. After that the village of Dresovka on the Don, some 25 km northeast of Ivanovka, they were surrounded by strong

Shurovka on October 17, 1942: After action with SS Infantry Regiment 8, the three armored cars of a platoon of the company are serviced and prepared for the next action. (AW)

October 19, 1942, near Smogilovka: The 2nd Police Armored Company lies in readiness. At left are three Renault tanks, at right three Steyr armored cars behind the village well. (AW)

Russian armored troops and motorized units at dawn on December 16, 1942, and fell into the hands of the Russians. Through too-early withdrawal of the Italian troops intended to defend the village, German formations (I./Police Regiment 6 and a Wehrmacht tank-destroyer unit) were surrounded and had to try to fight their way out alone.

On December 17, Police Regiment 14 was given offensive and defensive tasks between Ivanovka and Dresovka, but the situation was partly quite unclear because of the retreat of the Italian formations. Among others, the Commander of the III./14 fell, as did the Regimental Commander that evening, near Zapkova, ten kilometers east of Ivanovka, whither the regimental command post had been moved. The 14th (Armored) Company fought with their armored cars and tanks in the Novakalitva-Ivanovka-Zapkovo area and strengthened the 385th Infantry Division. The company's motorcycle platoons were used to deliver fuel to the police and Wehrmacht units.

On December 18 the I./14 was attacked again and again and had to withdraw to Ivanovka. The III./14 saw service with the 385th Infantry Division north of Ivanovka, near Novakalitva. On December 19 further hard fighting took place, so that the First Battalion had to give up Ivanovka in the evening, after heavy losses. By using all available vehicles of the 14th (Armored) Company and abandoning materials and baggage, they were able to take immobile wounded men with them when they evacuated Ivanovka and then to evacuate the main dressing station at Krinitchnaya.

By the evening of December 19, police regiment 14 down to the companies, and the 14th (Armored) Company down to the platoons, were apportioned to units of the 385th Infantry Division. One platoon of armored scout cars and tanks was used to secure the division staff of the 385th Infantry Division.

The 14th (Armored) Company had lost no men during its service at the front until December 21, 1942. The 1st, 2nd, 9th, 10th and 11th Companies (1 Platoon) of Police Regiment 14, on the other hand, had a strength of only 210 men, with only 70 rifles and five machine pistols, until December 22. Until December 23 the units of Police Regiment 14, which had been scattered repeatedly, were constantly reassembled and reformed for use as parts of the 385th Infantry Division.

After the defensive combat in the smaller Don bend, what remained of Police Regiment 14 was sent back to Mogilev. From there, as of March 5, 1943, the 14th (Armored) Company was transported back to Vienna, where it arrived on March 12. The company took up quarters in a small hotel in Vienna-Vorder-Hainbach. No documentation of the losses of materials during the Russian service from October 8, 1942 to March 5, 1943 have been found. But at least one Steyr armored car was burned by enemy fire and one Renault tank blown up by its crew. Of the company's 118 members, eight were killed and eight wounded during this period. Three policemen suffered from freezing and three were reported missing.

One Renault tank tows another tank away. In the background, the company's quarters at 90 Dniepr Street in Mogilev can be seen. (AW)

New Formation, Structure and Equipment

By command of the Reichsführer-Chief in the Ministry of the Interior on May 17, 1943 (O-Kdo I K (1b) Nr. 199/43), the company was reorganized at the Police Motor Vehicle School in Vienna.

The 2nd Police Armored company (new) was then composed of the following:

Leader Group with supply train and Workshop Platoon 11
1st Armored Car Platoon (3 Steyr armored cars)
2nd Armored Car Platoon (3 Steyr armored cars)
3rd Tank Platoon (5 Panzer I tanks).

The personnel of the company numbered 199 men, as an intelligence platoon leader had been added when the company was equipped with radios. The Panzer I tanks were the VK 1801 type.

The company was quartered in a hotel in Vienna-Vorder-Hainbach until June 19, 1943. Then from June 20 to July 3, 1943 the company was transferred for training to the Königsbrück Troop Training Camp near Dresden. The Indian Legion was there at the same time.

On July 4, 1943 the company returned from Königsbrück to Vienna-Vorder-Hainbach, where it remained until August 23, 1943.

Action

According to Order No. 3471 from the Chief of the Ordnungspolizei on August 10, 1943 (Kdo I g I a (1) Nr. 649/43 (g)), the 2nd Police Armored Company was subordinated to SS Police Regiment 4 and sent on the march from the Police Motor Vehicle School in Vienna to the Stanislau area for service against Russia. Nothing its known of its actions until mid-1944. Supervision of Ordnungspolizei forces and their actions as of July 15, 1944 was carried out by the company in the district of the HSSPF. Russia-Center and White Ruthenia, stationed in Schröttersburg, East Prussia.

As of May 1943 the 2nd Police Armored Company was reorganized and again included two armored scout platoons with three Steyr armored cars each. The six Steyrs were painted a dark yellow and equipped with radio

The company was named a regimental unit of SS Police Regiment 4.

In the district of the HSSPF Russia-Center and White Ruthenia (SS Group Leader and Lieutenant General of the police von Gottberg) there were at the same time also the strengthened Police Armored Companies 9 (Rest) and 12.

At this time Police Armored company 2. subordinated to SS Police Regiment 4, fought within the area of the "von Gottberg" Battle Group (Police Regiments 2, 4, 17 and 34) within General Command VI. Army Corps. On July 5, 1944 the war diary of General Command VI. Army Corps stated:

"5:00 A.M., situation: Attack of some 1000-man bands with grenade launchers and machine guns against railway line 15 km northwest of Ivie. Pol.Schtz.Rgt. 17 and 36 fight off the attack and push into the band with their police armored cars. Two antitank guns destroyed, numerous machine guns, grenade launchers and guns captured.

3:40 P.M.: Our police armored cars have shot down two Russian tanks."

Unfortunately, it cannot be told from the report which police company (2, 9 or 12) shot down the tanks.

In August (8/10-16/1944) SS Police Regiment 4 was with the 2nd Police Armored Company, subordinate to the Anhalt Brigade (SS Police Regiment 2 (new), SS Police Regiment 4, smaller units or remains of units), which saw front service in the area south and southeast of Grayevo. The Anhalt Brigade was assigned at this time to the 14th Infantry Division in the group of General Command VI Army Corps. The 2nd Police Armored Company had at this time a strength of three officers and 116 NCO and men, including the remainder of a motorcycle rifle platoon. The trench strength was two officers and 57 NCO and men. Three armored scout cars and two tanks were still on hand. By the end of August all the armored vehicles except one scout car had been lost.

Shortly after that the company had to be sent to Vienna for refreshing, for the report on Ordnungspolizei forces as of October 20, 1944 lists the 2nd Police Armored Company in WK XVII, the district of the BdO Vienna.

Nothing is known of any further action by this company.

All other motor vehicles, such as motorcycles, Stkw. 14 (left front) and Wkw., were painted yellow. The red diamond (tactical symbol for armored vehicles) on the vehicles was the company's symbol, to which the number 2 was later added at the upper left. (AW)

Three of the company's five Panzer I tanks are seen at the Königsbrück Troop Training Camp near Dresden in the summer of 1943. The most heavily armored vehicles were armed with only two machine guns. (AW)

Radziechov, February 2, 1944: The company's Opel Blitz 3-ton field kitchen truck. To make field kitchens mobile and simultaneously protected, they were often mounted on trucks and fitted with a rigid body. (AW)

Five of the company's Steyr armored cars in Busk on February 9, 1944. The scout cars bear winter camouflage paint (its good effect can be seen), and some have tire chains. The signposts point to the field police station and the command of Supply Battalion 349. (AW)

On February 11, 1944 the company passes this Type D Panther tank of the Wehrmacht, with turret number 113, near Zolkiev. (AW)

In the spring of 1944, two armored cars of the 2nd Police Armored Company have gotten stuck in the soft ground beside a paved road. Without help from heavy towing tractors the two armored cars could not get loose. (AW)

A tank of the 2nd Police Armored Company was captured by the Americans and sent to the USA. This picture was taken at the Aberdeen Proving Grounds on January 29, 1947. Unfortunately, the vehicle was scrapped later. (AW)

3rd Police Armored company

Establishment, Structure and Equipment

The establishment of the 3rd Police Armored Company took place on orders from the Reichsführer-Chief of July 23, 1942 (O-Kdo. I K (2) 251 Nr. 110/42) at the Police Motor Vehicle School in Vienna. The 3rd Police Armored company was made up of the following units:

Leader Group with supply train and workshop platoon
1st Armored Car Platoon (3 Steyr armored cars)
2nd Armored Car Platoon (3 Steyr armored cars)
3rd Tank Platoon (5 Renault tanks)

After its establishment, the 2nd Armored Car Platoon of the 3rd Police Armored company was ordered by the Reichsführer-Chief (O-Kdo. I K (2) 251 Nr. 110II/42), on August 4, 1942, to be turned over to the 2nd Police Armored Company, and was established anew according to the Reichsführer-Chief's order of August 18, 1942 (O-Kdo. I K (2) Nr. 110IV/42). For this the IdO Vienna turned three Steyr armored cars with all their weapons over to the Police Motor Vehicle School in Vienna.

All the other vehicles were assigned by special order of the Reichsführer-Chief. The forces needed to make up the unit were to arrive at the Police Motor Vehicle School, Landstrasser Hauptstrasse 68, Vienna, by 8:00 P.M. on August 27, 1942.

Action

After it was established, the 3rd Police Armored Company was sent to Police Regiment 15 of the HSSPF Russia-Center. Here the company saw service with the II./Police Regiment 15 in the Rokitno Swamps until early December 1942. At first it fought against small partisan bands in the area west of Pinsk, later against bands of more than 1000 men on the Sluzk east of Pinsk, and prevented the assembling of White Ruthenian and Volhynish partisan groups along the Pripyet.

On December 1, 1942 Police Regiment 15 returned to Pinsk with the 3rd Police Armored Company. Shortly thereafter. Police Regiment 15, including the 3rd Police Armored company, was loaded onto trains to return to service at the front. The troops passed through Gomel, Bakmach, Konotop, Kharkov and Kupyansk to Valuiki, where they were unloaded. On the way the first train, on which the Police Armored Company was loaded, hit a mine. One armored scout car fell off an open railroad car, and it took a day to replace it. After a week's stay, Police Regiment 15 marched by auto from Valuiki through Alexeyevka to Tatarino, where it was to take a position along the Don south of Voronesh with the Hungarian Army Corps. But it saw no action with the Hungarians.

In the night of December 17-18, 1942 there was an alarm, and Police Regiment 15 had to march with the 3rd Police Armored Company to Rossosh in bitter cold, on snow-covered roads and under great hardships. There the regiment learned that it was to be subordinated to the Fegelein Battle Group within the 24th Armored Corps. At the same time, there came the first news of the Russian break through the Italian positions with 1000 tanks, as well as the Russian advance along the Tshir to Millerovo north and south of Stalingrad.

In Rossosh the regiment met the first retreating Italians, but marched on toward the Don bend and took up alarm quarters in Michalovska.

On the evening of December 18 the march went on to Smagleyevskaya. The Regimental Commander, a battalion commander, the regiment's motor vehicle officer and a driver lost their way with their two cars and were captured by the Russians.

The II./Police Regiment 15 marched on in the Boguchar Valley, through Golaya to Bugayevka. Here the 3rd Police Armored company took on reconnaissance and securing tasks. In the darkness, Russian tanks rolled past the group in the direction of Kantemirovka. From Bugayevka the battalion, after coming through heavy Russian artillery fire, reached the Taly support point. The Führer had ordered that Taly had to be held until either reinforcements or German units already encircled by the Russians at Stalingrad or the Tshir reached it.

On December 20 and 21, 1942 heavy defensive fighting took place in "Igel Taly" against strong Russian troops. In the early morning hours of December 21 the evacuation of Taly was ordered.

The I./Police Regiment 15 was meanwhile fighting hard against a Russian rifle regiment and suffered heavy losses.

The battle group that broke out of Taly marched through Fisenkovo to Golaya and was attacked twice by Russian troops. Until December 24, 1942, heavy enemy attacks from the Goli Valley were repulsed, and police Regiment 15 and the 3rd Police Armored Company sustained heavy losses. More detailed information about the action of the 3rd Police Armored Company and Police Regiment 15 at the Don bend have not been found. It is known that the II./Police Regiment 15 lost more than 1000 dead, wounded and missing men.

Police Regiment 15, like Police Regiment 14 (see 2nd Police Armored Co.), was wiped out in the combat at the great Don bend between the Don and Donets in January 1943. The 3rd Police Armored company, subordinate to Police Regiment 15, was also wiped out or suffered heavy losses.

New Establishment, Structure and Equipment

To date, no information on the reestablishment of the 3rd Police Armored Company has been found, but it probably took place similarly to that of the 2nd Police Armored Company (see there). The structure and equipping were probably similar too, but this has not been documented:

Leader Group with supply train and workshop platoon
1st Armored Car Platoon (3 Steyr armored cars)
2nd Armored Car Platoon (3 Steyr armored cars)
3rd Tank Platoon (Five Panzer I tanks)

According to data from a former member of the company, he was transferred from his home base in Berlin to the Motor Vehicle School in Vienna on October 2, 1943 and ordered to join the 3rd Police Armored Company as armored platoon leader (tank platoon). At this time the tank platoon, or the whole company, was still being refreshed in Vienna. The equipment of the tank platoon is not known, but according to former members it could have been Panzer I (VK 1801) tanks. On October 5, 1943 a hasty loading for the eastern front took place, to join SS Police Regiment 16 (HSSPF, Russia-North) in the Nevel area. SS Police Regiment 16 stated in a report to the Commander of the Ordnungspolizei for the eastern lands, located in Riga, on October 20, 1943 that the strength of the 3rd Police Armored Company was four officers and 117 NCO and men.

Action

From November 1943 on, the company saw front service with the "Jeckeln" Battle Group of the VIII. Army Corps. Its action took place along with the Army's 32nd, 83rd and 263rd Infantry Divisions.

The war structure report of the "Jeckeln" Battle Group listed for Police Armored Companies 3 and 8, serving within it, an equipment of ten armored scout cars and six T26 tanks. On November 19, 1943 the VIII. Army Corps reported four light tanks and several armored scout cars in the "Jeckeln" Group.

The KTB of the 132nd Infantry Division reported on November 26, 1943 that Police Regiment 16 (with eight captured tanks) was subordinated to the 132nd Infantry Division.

On November 28, 1943 the KTB of the VIII. Army Corps recorded that on the following day the "Wagner" Battle Group (132nd I.D.) should approach the focal point from the Samoshye area and reach Vassilieva and Height 176.8. Subordinate police units (Police Regiment 16 and II./Police Regiment 9) were to take Height 192.7 to the south and the heights near Tukalovo and Puski from the west. The operation was supported by two light artillery units and a heavy battery. The captured tanks of the police were also available.

During the afternoon of November 29, 1943 the General Command reported the planned attack of the 132nd Infantry Division in the "Bekassine II" area. According to it, G.R. 426, strengthened by a fusilier battalion and an engineer battalion, and with two subordinate police battalions, supported by three light artillery units, one heavy howitzer battery and about eight captured tanks, was to move eastward from the Lopatovo area for the purpose of reaching the Shitnikovo Heights to close the Volosheno-Nevedro Lake narrows, with securing toward the Jasno Lake.

The war structure report of the "Jeckeln" Battle Group of December 1, 1943 listed a light armored company with eight tanks in Police Regiment 16, and that of December 11, 1943 included a normally structured police armored company with six light armored scout cars and five tanks.

The armored cars of the 3rd and 8th Police Armored Companies are not clearly identified in the reports of the VIII. Army Corps and the "Jeckeln" Battle Group, and it is not possible to identify them.

The "Jeckeln" Battle Group reported in their daily tank reports from November 19 to 30 four to ten light tanks or captured tanks, and from December 4 to 17 three to six Panzer I and six to nine light captured tanks, without telling which types they were. It can be presumed that they were the Panzer I tanks of the Tank Platoon of the 3rd Police Armored Company established in Vienna around the beginning of October, but this is not definite.

As of December 18, 1943, parts of the "Jeckeln" Battle Group were used against partisan bands within A.O.K. 16. On this subject, the HSSPF Eastlands and Russia-North reported to A.O.K. 16 on December 19, 1943 that Operation "Ottö, under his command, began on December 12, 1943. SS Police Regiment 16 (two battalions), SS Police Regiment 26 (three battalions), three alarm battalions, one alarm half-battalion, four foreigners' police battalions, two TN companies, one tank company, one scout car company, four flak-fighting troops and one light flak platoon were involved. To carry out Operation "Ottö, the surrounding line of Rudnya-Klyastizy-Zerkovichtshe-Borkovici-Martinovo-Strelki-Chernovoki-Stoki had to be reached and fought against in the southeast and west. The attack group was to form in the area on both sides of Sebesh and push forward southward as far as the Duenaburg-Polozk railroad line. It was thought that ten days would be needed for the operation. The war structure report of the "Jeckeln" Battle Group of December 21, 1943 listed an armored company with ten tanks as part of Police Regiment 16.

Until the middle of 1944, almost the only reports are those of a former company member. According to them, the company was almost completely wiped out in March 1944.

On June 17, 1944 the Chief of the Ordnungspolizei gave an order (Kdo. g I Org. (3) Nr. 121/44 (g.)) that SS Police Regiment 16 was to be reestablished quickly by the BdO, Riga. In the structure ordered for the new unit, the 15th company of the regiment was to be a tank company, to be structured as the 3rd Police Armored Company had been. Whether the 3rd Police Armored Company was thus refreshed and reformed is not known. From June to September 1944 the company saw service in the area between Vilna and Bauske and had only two armored scout cars. The others were used by the 215th Infantry Division in the end, though there are no definite data on this.

The review of the Ordnungspolizei forces and their service as of October 20, 1944 lists SS Police Regiment 16 with the 3rd Police Armored company in the territory of the BdO Eastlands, then in Liebau (HSSPF Eastland and Russia-North). According to reports of missing men, the 3rd Police Armored Company could have been in the Vienna-St. Poelten area for reforming and refreshing at the end of 1944 and beginning of 1945.

4th Police Armored Company

Establishment, Structure and Equipment
The order to establish the 4th Police Armored Company was given by the Reichsführer-Chief in the Ministry of the Interior on August 27, 1942 (O-Kdo. I K (2) 251 Nr. 157/42). According to it, the company to be formed at the Police Motor Vehicle School in Vienna was a formed in line with the strength and equipment directive of July 25, 1942 (see Appendix 5) as follows:

  Leader group with supply train and workshop platoon
  1st Armored Car Platoon (3 Steyr armored cars)
  2nd Armored Car Platoon (3 Steyr armored cars)
  3rd Tank Platoon (5 Hotchkiss tanks)

Old Steyr armored cars that were at the Vienna PV or Police Motor Vehicle School in Vienna were to be used for the armored car platoons. For this purpose the Vienna Police was to turn over the vehicles on hand there, with all their weapons, to the Police Motor Vehicle School. All remaining vehicles were to be distributed by special order. The personnel of the workshop platoon came from the Police School for Technology and Transit in Berlin during the formation period. This school was also to send a field kitchen to the Police Motor Vehicle School in Vienna for the armored company.

To form the two planned Steyr armored car platoons and the tank platoon, crews already trained by the Vienna PV were to be transferred from their homeland assignments to the 4th Armored Company at the Police

Motor Vehicle School in Vienna. Drivers, motorcyclists and other personnel were likewise to be transferred from various homeland units.

These forces were to be ready for service by September 10, 1942, so that their summoning to the Police Motor Vehicle School, Landstrasser Hauptstrasse 68, Vienna, could take place at any time.

Equipping the personnel with service clothing was to take place according to the appendix of a directive of July 12, 1941 (MBliV. p. 1303) and its subsequent additions. As for weapons, every policeman was to receive an 08 pistol with 80 bullets, or a 7.65 mm pistol with 50, and an 84/98 bayonet from his home base. The special clothing called for in Section M of Appendix 1 PBK1V., Part II, and a Type 30 gas mask with breathing tube were to be provided by the Vienna Police. In addition, the necessary 98a carbines with 150 bullets each were provided by the Vienna Sub-armory for the further equipping of the armored car platoons.

The members of the workshop platoon were to bring their rifles along from their home bases. The 4th Police Armored Company had a planned strength of 118 men (3 officers, 14 NCO, 35 men and 66 drivers). By order of the Reichsführer-Chief of September 12, 1942 (O-Kdo. I K (2) 251 Nr. 157 V/42) it was required that the four forces ready for the Armored Company were to be set in march at the right time so that they would reach the Police Motor Vehicle School in Vienna by 8:00 P.M. on September 23, 1942.

This command was overridden by that of September 18, 1942 (O-Kdo. I K (2) 251 Nr. 189/42), stating that the men were to be called in later by special instructions.

They were summoned by the Reichsführer-Chief's order of October 24, 1942 (O-Kdo. I K (2) 251 Nr. 227/42), by which the forces in readiness were to be sent on the march at the right time to arrive at the Police Motor Vehicle School in Vienna by 8:00 P.M. on October 28, 1942. According to the Reichsführer-Chief's order in the Ministry of the interior of November 3, 1942 (Nr. 101 -O-Kdo. I K (2) 207 Nr. 267/42), the company was assigned Workshop Platoon 13, formed at the Police School for Technology and Traffic in Berlin, with a strength of one master and seven patrolmen. Workshop Platoon 13 arrived at the company's school on November 14, 1942.

Action

According to the Reichsführer-Chief's order of November 26, 1942 (KdO. I-Ia (1) 1 Nr. 210/42), the 4th Police Armored Company was subordinated to Police Regiment 13, which was in the territory of the HSSPF. Russia-Center. The company set out on the march to Police Regiment 13 on December 18, 1942. At present, no records of their actions for the first half of 1943 have been found.

On June 17, 1943 and January 15, 1944, Police Regiment 13 with its subordinated Police Armored Company 4 have been documented as being in the area of the HSSPF. Russia-Center and White Ruthenia. Armored Company 4 took part in various actions against partisans.

A Steyr armored car of the 4th Police Armored Company is seen in the Russia-Center zone in the winter of 1942-43. The vehicle has been painted white with chalk fluid. This type of winter camouflage was chosen because it could be washed off. (BAK)

On January 10, 1944 the First Army Corps took command of all troops and bands in the Polozk-Jasno Lake South End sector. These included the "von Gottberg" Police Battle Group. In a structure report of this battle group of January 10, 1944, the company was listed as part of Police Regiment 13, with a strength of five armored cars and five tracked armored vehicles. On this day its fighting strength amounted to one officer and 51 men, with a supply strength of one officer and 71 men. The company now had only one third of its specified strength. In a report of the "von Gottberg" Battle Group on March 1, 1944, the equipment was listed again as five armored cars and five tracked armored vehicles.

On March 25, 1944 the First Army Corps reported that the 4th Police Armored Company had left its area.

Between March and June 1944 the company was transferred from SS Police Regiment 13 in Russia to the Eastlands, in the territory of the BdO Salzburg (WK XVIII). Here too, its action consisted of fighting against partisans.

The overview of Ordnungspolizei forces and their action as of July 15, 1944 lists the company in the area of the BdO Salzburg. The company was named as a unit of SS Police Regiment 13, based at St. Jakob i.R. An assignment of replacement forces by the Police Armored Replacement Unit on August 29, 1944 lists the 4th Police Armored Company's quarters as being in Ferlach near Klagenfurt.

The overview of Ordnungspolizei forces and their actions as of October 20, 1944 also lists the 4th Police Armored Company as being with SS Police Regiment 13 in the area of the BdO Salzburg, HSSPF Alpinelands.

The winter camouflage with chalk water was applied only to the combat vehicles at first. The gray truck and the men in green coats stand out clearly from the white surroundings. (BAK)

The Steyr armored cars were armed with a 2 cm tank gun in the turret and a water-cooled 7.92 mm Schwarzlose machine gun. Additional air-cooled 7.92 mm machine guns were installed, one at the bow and the other at the rear. (BAK)

A Hotchkiss tank of the 4th Police Armored Company is seen outside a burning village in the winter of 1942-43. In fighting against partisans, localities that were suspected of being partisan support points were burned down completely. (BAK)

The Hotchkiss tanks were armed with a 3.7 cm tank gun and a 7.92 mm machine gun. The commander had to operate both weapons himself. (BAK)

All Hotchkiss tanks taken over by the Wehrmacht were fitted with a rear spur. This support improved the tanks' ditch-spanning capability. (BAK)

In rough terrain it was not easy, especially in winter, to steer the armored vehicles. Uneven ground and holes were hidden by the snow. (BAK)

It was unfortunate that a ditch or a lake with thin ice was sometimes hidden under the snow. Here a Hotchkiss of the 3th Police Armored company has broken through the ice into a water hole. A rescue attempt is being made by a pair of other Hotchkiss tanks. (JB)

Summer 1943: The field kitchen of the 4th Police Armored Company, with field kitchen and water tank trailer and other utensils, is shown. The truck at right is a 3-ton Ford, probably used as Stkw. 14. (JB)
It was not only in winter that a tank could get stuck. There were good opportunities in the summer too. Especially in these situations, the lack of suitable towing tractors in the armored companies was very obvious. Often only the use of the company's other tanks helped. (KF)

5th Police Armored company

Establishment, Structure and Equipment
The 5th Police Armored company was established on orders from the Reichsführer-Chief (O-Kdo. I K (2) 251 Nr. 192/42, of September 22, 1942, and O-Kdo. I K (2) 251 Nr. 225/42, of October 21, 1942) at the Police Motor Vehicle School in Vienna, and was structured according to the Equipment and Strength Instructions (see Appendix 5) as follows:

Leader Group including Supply Train and Workshop Platoon
1st Armored Car platoon (3 Steyr armored cars)
2nd Armored Car Platoon (3 Steyr armored cars)
3rd Tank Platoon (5 Hotchkiss tanks)
The company was assigned Workshop Platoon 14, with a strength of one master and seven patrolmen.

Action
From December 1942 on, the company was fighting partisan bands in Russia, where it was subordinated to Police Regiment 10 in the area of the HSSPF Russia-South. One of the company's locations was Novograd-Volynsk on the Slutch.

In the period from December 15, 1942 to the beginning of 1944 it took part in anti-partisan fighting and combat with stronger groups in the regions of Dvruch-Salizovka-Gorodniza-Shitomir-Kodra-Sabutanye-Korostichev-Iviniza-Luzk. In combat in Novy-Zaharov on September 11, 1943 one officer was wounded.

At the beginning of 1944 the company was withdrawn to Poland (Galicia) for refreshing, along with SS Police Regiment 10. There it was stationed in the Poniativa camp from February through the end of April 1944.

Restructuring to a Strengthened Police Armored company
With a message from the Reichsführer-Chief on March 2, 1944 (O-Kdo. In K (2) 251 E Nr. 104/44), the restructuring of the 5th Police Armored company into a strengthened Police Armored Company was ordered. According to the Strength and Equipment records, the company was structured as follows:

Leader Group including intelligence platoon, workshop platoon and supply train
1st Armored Car Platoon (3 Steyr armored cars)
2nd Tank Platoon (5 Hotchkiss tanks)
3rd Tank Platoon (5 T34 tanks)
4th Tank Platoon (5 T34 tanks)
The needed personnel were to be summoned from the Police Motor Vehicle School in Vienna (Police Armored Replacement Unit). Training of the added personnel by the Police Armored Replacement Unit was not to take place, and they had rather to be trained by the 5th (reinforced) Police Armored Company itself.

The ten T 34 tanks for the 3rd and 4th Platoons were already at hand in the company. All the other vehicles were to be provided by the Police Motor Vehicle School in Vienna from those it had on hand. Three Steyr armored cars, one Stkw. 4 and one motorcycle were returned to the Police Motor Vehicle School in Vienna by the 5th (reinforced) Police Armored Company.

The lacking personnel for Workshop Platoon 14 of the 5th (reinforced) Police Armored Company were to be provided from the men assigned for training to the Police Motor Vehicle School in Vienna (Armored Replacement Unit). As a further strengthening of the 5th (reinforced) Police Armored company, one officer, one shoemaker and one tailor were assigned. The forces assigned for strengthening had to be born in or after 1901, and had to arrive at the Police Motor Vehicle School, Landstrasser Hauptstrasse 68, Vienna, by 8:00 P.M. on March 14, 1944.

As for clothing, the home bases were to provide a field cap, a field shirt of older or newer type, long trousers, a coat, a pair of heavy laced shoes, a pair of snap leggings, a scarf or collar tie, three undershirts, three pairs of shorts, three pairs of socks and a belt with pocket and buckle.

For further equipping with service clothing, Appendix 1 – for Police Vehicle Crews, Appendix 1a – from a memo of July 12, 1941 (MBliV.S. 1303), expanded on by memos of August 5 and December 15, 1942

(MBliV.S.1644 and MBliV.S. 2538). The lacking (not brought along) service clothing had to be provided by the Police Motor Vehicle School in Vienna (Police Armored Replacement Unit) (compare Section B of Memo MBliV. S. 2338 of December 15, 1942). The Vienna PV likewise had to supply a Gas Mack 30 with breathing tube.

Home bases were to provide a Pistol 08 or 7.65 with 80 or 50 bullets respectively, and an 84/98 bayonet as weapons.

Action

The 5th (Strengthened) Police Armored Company remained subordinate to SS Police Regiment 10 until mid-1944, seeing service with it in the area of the Hst.SSPF Ukraine. From April 30 to June 13, 1944 the company, as part of the "Pruetzmann" Battle Group, took part in front service near Stanislavcyl in the Brody area.

On June 23, 1944 the Chief of the Ordnungspolizei sent out an order (Kdo. g I-Ia (1) Nr. 178/44 (g)) to the Hst.SSPF of the Ukraine (BdO) in Tarnov, the HöhSSPF Italy (BdO) in Verona and the HSSPF Adriatic Coast (BdO) in Trieste. On orders from the RFSS of June 21, 1944, SS Police Regiment 10 was immediately to transfer with all regimental units, including Police Intelligence Company 121, officers, NCOs and men, and all their equipment, vehicles and supplies. SS Police Regiment 10 was sent by rail to Gorizia and subordinated to the BdO Trieste. A special order was to follow as to the regiment's and regimental units' refreshing and reequipping for mountain warfare. The 5th (strengthened) Police Armored company was thus transferred to Italy with SS Police Regiment 10.

In a message of July 7, 1944 (Kdo. g I Org./Ia (3) Nr. 178/44 (g)), the Chief of the Ordnungspolizei ordered the reorganization of SS Police Regiment 10. The regiment's reorganization essentially consisted of enlarging the battalion from three to four companies. One armored company was to become part of the regiment as the 15th Company according to the previous structure of the 3rd Police Armored Company. The strengthened 5th Police Armored Company, though, remained with the regiment by the structuring ordered on March 2, 1944.

Overseeing the Ordnungspolizei forces and their actions, with the status of July 15, 1944, was done by the company as a regimental unit of SS Police Regiment 10 in the area of the BdO Trieste (HSSPF Adriatic Coast), stationed at Aidurana-Vipacco, Gorizia. Overseeing on the October 20, 1944 basis brought no changes from that of July 15, and the company remained stationed in Gorizia.

The 5th (strengthened) Police Armored Company took part in the following actions in Italy:

10/8 to 10/22/1944: Operation "Waldlaeufer" in the Tolmezzo-Paluzza-Villa-Santina area.

10/25-11/1/1944: Securing ammunition transport in the Gorizia-Aidussina area.

11/27/1944: Combat with strong partisan bands in the Sambasso area.

12/19-12/23/1944: Operation "Adler" in the Gorizia-Sambasso-Vipacco area.

1/19-1/21/1945: Action in the Raunizza-Zagorie-Britovo area.

On January 19, 1945 a T-34 platoon received an order to leave Gorizia and relieve the "Fulmine" Battalion of the "Flottiglia-Mas" Division, which was surrounded by strong partisan bands in Tarnova. A shock troop, consisting of the T-34 tank platoon, an engineer platoon and forty men of the "Flottiglia-Mas" as infantry, advanced via Salcano-Raunizza. The advance took part mostly at night in a heavy snowfall. The roads were heavily mined and under the heaviest grenade-launcher and machine-gun fire. For two days the shock troop tried to advance on the well-constructed and camouflaged positions of the bandits, who were far superior in men and weapons, in the hills north and west of Raunizza. Only on the third day, after the shock troop had been strengthened by the III./SS Police Regiment 15, did the attack attain the heights on three sides in cooperation with the III./SS Police Regiment 10. Only after six hours of heavy, ceaseless firefights by the tanks could the heights be taken by the infantry. The partisans lost 92 dead and three to four times that many injured. The shock troop lost seventeen wounded.

The last evidence of action by the 5th (reinforced) Police Armored company comes from a code-name list of the intelligence leader in the area of the HSSPF Adriatic Coast, dated February 21, 1945. In this list, valid as of midnight on March 5, 1945, the company was listed under the code name "Backpflaume" (baking plum).

A Hotchkiss tank (RMdI.-Nr. 196 25) of the 5th Police Armored company at a support point in the Shitomir area in the spring of 1943. The Hotchkiss does not yet have a radio built in; the bracket for the antenna base at the rear of the right track apron is empty. The troops supplied themselves from the land; here they are getting milk. (PFA)

Cattle served not only for milk production but were also used as draft animals, as here to transport wood. In front of the building is an Stkw. 4, with a Steyr armored car visible inside. The windows of the building are taped to prevent splinters from flying if shots break the glass. (JF)

A Steyr armored car is prepared for action. It is a first-series (1935-37) vehicle, once used by the Austrian Gendarmerie. Only the Gendarmerie vehicles were fitted with lights on top of the turret. The mixed uniforms of the policemen are also interesting. (JF)

This Steyr armored car is joined for action by a truck and a Hotchkiss tank. Such partial units were used, for example, for patrolling in unoccupied areas or to escort columns of vehicles through partisan areas. (JF)

Above: The armored vehicles are accompanied in action by a three-ton truck carrying riflemen. Only limited seating for the riflemen is available in the back of the truck. In this photo too, the policemen's varied uniforms are noteworthy. (JF)

Left: Two Hotchkiss tanks of the company are seen in Poland in the spring of 1944. Both are painted with a new yellow camouflage paint. After it was applied, the German crosses, which used to be at the right front of the tanks' bow plates, were painted in the center of the front. Thus the Hotchkiss tanks could be recognized more easily than German tanks. (JF)

The Poniatova camp in Poland, seen early in 1944, is surrounded by a high wire fence. A sign with the field post number of the 5th strengthened Police Armored Company indicates what unit is here. At the left front is a T34 tank of the company. (JF)

The crew of a T34 tank platoon stands at attention in the yard of the camp. Again one notes the varied uniform parts worn by the men, perhaps based on their various tasks. Some of the tank crews (commanders and drivers?) are wearing SS camouflage jackets and dust-protector goggles. (JF)

Vehicles of a T34 platoon are seen in Poniatova in 1944. In front are two 1943 model T34s, recognizable by the larger turret with two round hatches; behind them is a 1941 model T34 with a smaller turret and one large hatch. All T34 tanks had German crosses painted on the bow, turret sides and top. (JF)

In the Poniatova camp in Poland, this police tank platoon's three heavy Russian scout cars were photographed along with the reinforced 5th Police Armored company. It is not known whether the tank platoon was assigned or subordinated to the reinforced 5th Police Armored Company. (JF)

Above: At Gorizia in the summer of 1944, the commander of the forward Hotchkiss tank tests the speaker system of his vehicle. Behind the Hotchkiss are a BA 64 and a Steyr armored car. The small ex-Russian BA 64 was already with the company in Poniatova. (JF)

Left: After their transfer to the Gorizia area, the vehicles of the 5th reinforced Police Armored Company were camouflage-painted anew in the summer of 1944. Here on a T34 (1942 model) tank the German crosses have been newly painted in white; the black bars were added in a second work process. (JF)

Above: In the Gorizia area an armored Wehrmacht unit was also stationed in passing. It was equipped, among others, with Panzer III Type N tanks (short 7.5 cm tank guns). On August 4, 1944 the Wehrmacht tank soldiers and those of the reinforced 5th Police Armored Company gathered for this photo. (JF)

Left: This Hotchkiss tank's camouflage paint has also been redone or changed. The crosses are still lacking. The riveted turret in the left background belongs to an AB 41 armored car. These ex-Italian scout cars were used by the police armored troops, but not by the 5th reinforced Police Armored Company. (JF)

Two T34 tanks and a Steyr armored car are seen in the Gorizia area at the end of August 1944. In front is a 1943 model T34, behind it a 1941 model. The policemen's varied uniforms are of interest. Many camouflage jackets and trousers were made of Italian camouflage cloth. (AWB)

6th Police Armored Company

Establishment, Structure and Equipment

The 6th Police Armored Company, by order of the Reichsführer-Chief on October 28, 1942 (O-Kdo. I K (2) 251 Nr. 234/42), was established at the Police Motor Vehicle School in Vienna and structured as follows:

> Leader group including supply train and workshop platoon
> 1st armored car platoon (3 Steyr armored cars)
> 2nd armored car platoon (3 Steyr armored cars)
> 3rd tank platoon (5 Hotchkiss tanks)

The intended manpower was to be made ready for service by home bases so that they could be summoned by wire to the Police Motor Vehicle School, Landstrasser Hauptstrasse 68, Vienna, at any time. Before being sent on the march, they were to be tested by their home bases to see whether they were healthy enough to stand the strains put on armored-vehicle crews. Their equipping with service clothing was to be done according to Appendix I of an order of July 12, 1941 (MBliV. S. 1303), amplified by an order of August 5, 1942 (MBliV. S. 1644). The prescribed special clothing, including a black coat, long black trousers, a black field (fore-and-aft) cap, and a combination suit of denim, as well as a Gas Mask 30 with breathing tube, were to be provided by the Police School in Vienna. Instead of a crash helmet, they received steel helmets. The home bases were to provide each man with a Pistol 08 or 7.65 mm with 80 or 50 bullets, and an 84/98 bayonet.

When ready, the forces were to be sent on the march at the right time, according to the Reichsführer-Chief's order of March 30, 1943 (O-Kdo. I K (1b) 251 V/43), so that they would arrive at the Police Motor Vehicle School, Landstrasser Hauptstrasse 68, Vienna, by 8:00 P.M. on April 5, 1943.

The 6th Police Armored Company was assigned, according to an order of the Reichsführer-Chief in the Ministry of the Interior on 4/21/1943 (O.Kdo I K (1b) 207 Nr. 105/43), Workshop Platoon 26, which was established at the Police School for Technology and Traffic in Berlin, with one master as workshop leader and seven NCO as skilled workers. The Workshop Platoon joined the Company at the school in Vienna on May 9, 1943.

In November 1943 the Company was expanded by a motorcycle rifle platoon (strength 2/24), about the establishment and assignment of which no data are at hand. According to information from company members, the motorcycle rifle platoon also had a group with light grenade launchers.

On June 3, 1944 the Chief of the Ordnungspolizei ordered (Kdo. g I Org. (3) Nr. 109/44 (g)) the establishment of an engineer group (mine-detecting troop) composed of one NCO and ten men for the 6th Police Armored Company. Accordingly, this engineer group was to be set up immediately and trained by the Police Weapons School I in Dresden-Hellerau for the 6th Police Armored Company, then serving in the area of the BdO. Croatia. The engineer group was intended primarily for mine-detecting and -removal duties. The personnel needed for the engineer group were to be provided by Police Weapons School I. 50% of the personnel were to come from the German Reich. Recuperating NCOs who were not capable of being group leaders could be included. To equip the engineer group, the Police School of Technology and Traffic – Main Arsenal in Berlin was to provide Police Weapons School I with one MG 26(t) with equipment and 1500 bullets, one MP 40 with equipment and 500 bullets, nine Carbine 98k with equipment and 120 bullets each, one 7.65 mm pistol with equipment and 50 bullets, one flare pistol with ten white, red and green cartridges, eleven 84/98 bayonets, one 6 x 30 telescope, one marching compass, one mine-detecting device with equipment, two short or folding spades, one small wire cutter, two half-size spades, two half-size pickaxes, one half-size axe, two hatchets, five Cleaning Device 34, eleven complete gas masks, and five complete flashlights.

The conclusion of the group's training was to be reported to the Chief of the Ordnungspolizei by Police Weapons School I by wire. Marching to the BdO. Croatia was to follow by special order. Three months after the arrival of the engineer group, the BdO. Croatia was to file an unsolicited report about the experience gained by this group. Motor vehicles were not specifically assigned to the engineer group (mine-detecting troop).

Police Weapons School I (PWS I) in Dresden-Hellerau reported by wire on June 14, 1944 that the engineer group's training was finished. The chief of the Ordnungspolizei gave the command, on June 19, 1944 (Kdo. g I-Ia (1) Nr. 176/44 (g)), for the engineer group to march to the BdO. Croatia, with transport readiness as of July 1, 1944.

Action

Based on a command from the Chief of the Ordnungspolizei of June 16, 1943 (Kdo Lg Ia (1) Nr. 144/43 (g)), the 6th Police Armored company was transferred to outside service in Croatia and subordinated to the BdO. Esseg. A pre-command of one officer and five NCOs had already been sent on the march on June 18, 1943.

As of September 16, 1943, under the command of SS Group Leader Generalleutnant der Polizei Kammerhofer, an operation to mop up the Fruska Gora in the Save bend was begun. The 6th Police Panzer company also took part in this operation. According to reconnaissance, a partisan brigade of some 1000 men was in that area, armed with two antitank guns, two grenade launchers, eight to ten heavy and some 125 light machine guns.

The assembled police forces (three battalions, one armored company, one SS special command) had the task, along with Wehrmacht units (tank destroyers, artillery, railroad securing), of first freeing the Syrmia area from partisans so that the railroad lines and main roads of the area were again usable without great danger and the ethnic German settlements appeared to be safe. In carrying out this pacification, police posts were to be set up as planned and responsible mayors were to be installed in the area of action. It was intended to carry out actions against individual localities or partisan groups first, and only to liberate larger areas later. The first operation set out from Ruma, directed against the higher partisan staff at the Villa Reinbrecht in Prnyavor and the village of Besenovo-Prnyavor, and took place on September 19, 1943.

The operation seems to have taken some time, for on October 4, 1943 the 173rd Reserve Division reported that the Kammerhofer police operation at the Save bend was being continued. The horse drawn police battalions scarcely made contact with the enemy. One motorized battle group, consisting of a police armored company, a battery and an engineer company of the 173rd Res. Div., advanced to Save-Vend Kupinski Kat (40 kilometers southwest of Belgrade). Weak partisan forces fled into the swamps, taking along their wounded and their dead men's weapons. Three partisan camps were wiped out.

Parts of the 6th Police Armored company (two armored scout cars and a motorcycle rifle platoon) took part in a KdO. Esseg action against the Serbian village of Koprivna on October 30, 1943. This was a reprisal measure for the murder of an SS man.

Likewise, parts of the 6th Police Armored Company took part in an attempted relief action on the town of Virovitica, which was surrounded by strong bandit forces. The relief forces used there (four police and three Ustascha companies) were accompanied by two armored police vehicles. While the relief forces were in combat 4 kilometers west-northwest of Podrav Slatna, both armored vehicles were put out of action.

In December 1943 the 6th Police Armored Company was prepared for Operation "Cannae". No precise data on this operation are available.

On May 25, 1944 the General Command, LXIX. Army Corps, gave the command for Operation "Schach" (Chess). The goal of the operation was to push the partisan bands in the Vojnic-Barilovic-Korana-Bach to Bukovac-Miholsysko area together and wipe them out. From May 26 to 28, 1944 parts of the 6th Police Armored Company (four 12-ton tanks) took part in this operation along with the strengthened Police Volunteer Battalion 7.

In July 1944, with the establishment of Police Volunteer Regiment 1 "Croatia", the 6th Police Armored Company had to be assigned to it as its armored company, for the ovewview of the Ordunugspolizei forces and actions as of July 15, 1944 lists the 6th Police Armored Company with a mine-detecting troop as a regimental unit of Police Volunteer Regiment 1 "Croatia". The company saw service at Djakovo in the area of the BdO Esseg.

The overview of Ordnungspolizei forces and their service as of October 20, 1944 includes the same subordination situation. The 6th Police Armored Company with its mine-detecting troop was reported as being in reserve at Syrmia in the BdO Esseg area, HSSPF Croatia at that time. At the same time as the 6th Police Armored Company, the 11th and 16th Police Armored Companies, each with its own mine-detecting troop, were also serving in the BdO Esseg area (see 11th and 16th Police Armored companies).

A troop structure chart of the General Command, XXXIV. Army Corps, as of December 14, 1944 lists the 6th Police Armored Company with parts (4 7.5 cm antitank guns (?)) with the Stephan Division z.b.V. (a unit made up of police, SS and Wehrmacht troops), and with its mass in the Police Security District South.

A list of the Stephan Division states that the parts of the 6th Police Armored Corps serving with it were stationed in the town of Nasice. These parts, within the Stephan Division, were assigned to a Police Security District North.

According to a war structure report from the Stephan Division on December 23, 1944, most of the company seems to have been subordinated to the 11th Luftwaffe Field Division, which had been assigned to the XXXIV. Army Corps in December 1944. On March 1, 1945 the 6th Police Armored Company was named in a message from the Intelligence Leader of the BdO. Croatia to the Intelligence Officer of the Police Security Region Drau, giving the company's radio plans.

Further information on the company could be gained from former members, but errors of time and place could result from the long lapse of time.

The company was apparently stationed at Osijek (Esseg), with its area of action extending from the Zagreb-Virovitica-Daruvar-Nasice district westward to Belgrade and Novi Sad.

The two scout-car platoons were in action almost constantly, the tank platoon more rarely. It is also reported that armored scout cars struck mines or were even shot down and crews wounded. Till the war ended, the company was subordinated to Luftwaffe and Army units. After the armistice of May 8, 1945, all the armored cars but two scout cars were abandoned in the area north of Zagreb. With the two scout cars and other vehicles the company marched via Celje, Lavamund and Voelkermark to St. Veith on the Glan, where their arms were laid down and the scout cars were abandoned. In Voelkermark they made their first contact with British troops. From St. Veith they marched in their vehicles to St. Michael-Tamsweg/Mauterndorf, where the company was interned by American troops.

7th Police Armored Company

Establishment, Structure and Equipment

The 7th Police Armored Company was established at the Police Motor Vehicle School in Vienna following an order from the Reichsführer-Chief on January 13, 1943 (O-Kdo. I K (2) 251 Nr. 11/43), and was structured as follows:

 Leader Group including supply train and workshop platoon
 1st Armored Car Platoon (3 Steyr armored cars)
 2nd Armored Car Platoon (3 Steyr armored cars)
 3rd Tank Platoon (5 Renault tanks)

 By Order No. 3267 of the Reichsführer-Chief of the German Police in the Ministry of the Interior, dated February 15, 1943 (O-Kdo. I K (2) 205 Nr. 10/43) it was ordained that the two Steyr armored car platoons with all their personnel and vehicles were assigned to the 13th (reinforced) Police Armored Company with the Griese SS Police Regiment.

 On February 16, 1943 the Reichsführer-Chief, in Message 793 (O-Kdo. I K (2) 205 Nr. 10II/43), ordered that two Panhard scout car platoons of the 8th Police Armored Company, with their personnel and vehicles, be assigned to the 7th Police Armored Company.

 The company was then structured as follows:
 Leader Group including supply train and workshop platoon
 1st Armored Car Platoon (3 Panhard armored cars)
 2nd Armored Car Platoon (3 Panhard armored cars)
 3rd Tank Platoon (5 Renault tanks)
 The company was assigned Workshop Platoon 6.

Action

By Order No. 4157 of the Chief of the Ordnungspolizei (Kdo. I-Ia (1) I Nr. 238/43) on March 18, 1943, the company was subordinated to SS Police Regiment 11. After leaving the Police Motor Vehicle School in Vienna, it reached the II./SS Police Regiment 11, in the area of the HSSPF Russia-South on March 29, 1943. An advance command of one officer and five NCOs had already been sent on the march on March 23.

 The company then saw service in 1943 and 1944 with the II./SS Police Regiment 11 in the area of the HSSPF Russia-South , in the Gomel district among others. At the same time, the 10th Police Armored Company was also in service with SS Police Regiment 11, but with the First Battalion. It appears from this that the two companies' places and areas of action sometimes equaled or overlapped each other (see also under 10th Police Armored Company).

 Both Police Armored Companies were transferred with SS Police Regiment 11 to strengthen the defensive forces in the big bend of the Vistula, seeing front service with the XIII. Army Corps. According to a message from the Chief of the Ordnungspolizei of September 13, 1944 (Kdo. In K (1b) 251 E Nr. 204/44), the 7th Police Armored Company, with all its personnel and equipment, was withdrawn from combat service with SS Police Regiment 11. It left the II./SS Police Regiment 11 on September 25 and arrived back at the Police Motor Vehicle School in Vienna on October 2, 1944.

 The battalion commander of the II./SS Police Regiment 11 reported the departure of the company in his daily battalion command on September 25, 1944:

  "The 7th Police Armored Company departed from the II./SS Police Regiment 11 on 9/25/1944. I take this occasion to thank all the members of the Armored Company for their loyal, noble fulfillment of their duties and exemplary behavior. The 7th Police Armored Company has always done its duty in partisan and front-line combat, whether with tanks or infantry action. It suffered the loss of 16 dead, including two officers, two missing and 44 wounded, which indicated the severity of the combat in which it took part. The Armored Company played a significant role in the success which the battalion achieved."

Refreshing

On July 7, 1944 the Chief of the Ordnungspolizei had ordered (Kdo. g I Org./Ia (3) Nr. 179/44 (g.))the reorganization of SS Police Regiment 11. According to this order, SS Police Regiment 11 was to be reorganized by the Commander of the Ordnungspolizei in Krakau. The Police Armored Companies then with the regiment (7th and 10th Companies) were to march to the Police Motor Vehicle School in Vienna with all their vehicles, weapons and equipment, for refreshment of their existing strength. In the outline of Ordnungspolizei forces and their actions as of July 15, 1944, the 7th Police Armored company was already reported from WK Zone XVII of the BdO. Vienna reported that the company arrived there, as noted above, on October 2, 1944.

The officers, NCOs and men of the company were, except the workshop platoon, assigned to the Police Armored Replacement Unit at the Police Motor Vehicle School in Vienna until ordered to march again. Company men wounded in action, ill or recovered were to be sent from their homeland billets to the Police Motor Vehicle School in Vienna.

The outline of Ordnungspolizei forces and their actions as of October 20, 1944 lists the 7th Police Armored Company in WK Zone XVII of the BdO Vienna. No information about completed refreshing of personnel, weapons or equipment is available.

Workshop Platoon 6, by order of the Reichsführer-Chief of March 3, 1944 (O-Kdo. In.K (1b) 202 E Nr. 245/44), had already been sent to the Workshop Replacement Unit at the Police Motor Vehicle School in Iglau, and after overhauling its vehicles and equipment, was again ready for outside service on September 26, 1944.

The 7th Police Armored Company's quarters in southern Russia in 1943. (AB)

A bed of straw and a place to relax in the open air. In the foreground is a Russian 7.92 mm Degtyarev machine gun with an ammunition drum. Spare drums were carried in boxes. (AB)

Action

Documentation of subsequent service of the 7th Police Armored Company is not available. A former member of the company mentions action on the Austro-Hungarian border at the war's end, but without armored vehicles. This may have been the last action of the Police Armored Replacement Unit.

Armored cars and other vehicles of the company's two scout-car platoons are parked here. Next to the Panhards, one of which is not seen, are the company's Stkw. 14 and Wkw. (KS)

An armored car platoon uses a sand causeway to advance through a wooded region. The leading Panhard bears German registration number 271 77. Behind it are the platoon's unarmored vehicles. (KS)

A destroyed wooden bridge compels the vehicles to leave the sand causeway. In the soft ground beside the causeway the vehicles sank up to their undertrays. Even the armored cars with their high wheels got stuck and required pieces of wood for support. (KS)

Only the tracked vehicles were still mobile in this terrain and could provide help. The Renaults were equipped with radios but still had the original turret cupola without hatches. (KS)

Above: Men of the 7th Police Armored Company are seen with a 7.5 cm self-propelled antitank gun on Gw. II of the SS Cavalry Division "Florian Geyer". While the police were equipped with captured small, weak Renault tanks, the Wehrmacht converted their Panzer II tanks to tank destroyers by mounting antitank guns on them. The "Florian Geyer" SS Division saw action with the 7th Police Armored Company in the summer of 1943. (KS)

Left: The Panhard armored car with registration number 271 77 is probably a platoon leader's car. The commander uses a signal stick to give orders to the platoon. The V-shaped plate on the bow was added by the Wehrmacht to afford greater shot protection. (BAK)

The commander of a Panhard armored car observes the effects of armored-car of artillery fire from the opened turret hatch. The hatch cover, opening to the front, thus protected the officer from frontal fire during observation. Depending on the situation, observing could also be done through the space between the hatch cover and the turret top, the periscope visible on the turret top, or the loopholes in the turret. (BAK)

8th Police Armored Company

Establishment, Structure and Equipment

According to an order from the Reichsführer-Chief in the Ministry of the Interior on January 22, 1943 (O-Kdo. I K (2) 251 Nr. 28/43), forces were to be kept ready for summoning to form armored companies as of February 13, 1943. According to the Reichsführer-Chief's order of February 3, 1943 (O-Kdo. I K (2) 251 Nr. 40/43), the forces held ready by the January 22, 1943 order were to be sent on the march to the Police Motor Vehicle School, Landstrasser Hauptstrasse 68, Vienna in time to arrive there by 8:00 P.M. on February 15, 1943 to form the 8th Police Armored Company. A structure of the 8th Police Armored Company has not been found to date, but it is probable that this company too was structured as follows:

Leader Group including supply train and workshop platoon
1st Armored Car Platoon (3 Panhard armored cars)
2nd Armored Car Platoon (3 Panhard armored cars)
3rd Tank Platoon (5 ?? tanks)

On February 16, 1942 the Reichsführer-Chief, in Order No. 793 (O-Kdo. I K (2) 205 Nr. 10II/43), had two Panhard armored scout car platoons of the 7th Police Armored Company, with their crews and vehicles, assigned to the planned 8th Police Armored Company.

No documentation of the equipping of the 8th Police Armored Company has been found to date, but it is shown by photos that the armored car platoons were equipped, like those of the 7th and 9th Companies, with Panhard armored cars. The tank platoon was probably supplied with captured Russian tanks, perhaps only some time after the company was transferred for outside service. In one photo a Panzer I (Type A) can also be seen, so that equipping with this type also seems possible.

In the schematic war structure report of the "Jeckeln" Battle Group, the 8th Police Armored Company is listed on December 1, 1943 as having ten armored vehicles, on December 11, 1943 with six armored cars and five tanks (representing the normal numbers), and on December 21, 1943 with twelve armored vehicles. Such a fast change in equipment would be possible if the 8th Police Armored Company had exchanged armored car platoons with the 3rd Police Armored Company, which also served with the "Jeckeln" Battle Group, but nothing is known of this.

Action

After its establishment, the company went into service under the HSSPF Eastlands and Russia-North. In the November 1943-March 1944 period, the company is documented as an independent unit of the VIII. Army Corps with the "Jeckeln" Group (a battle group of Police, SS and Army units, under the command of SS Obergruppenführer and General of the Police Jeckeln). At this time the battle group fought in the Nevedro- and Gussino Lake area.

In December 1943 the 8th Police Armored Company, as part of the "Möller" Securing Group (Commander: SS Oberführer Möller), took part in Operation "Ottö, beginning on December 20. This operation, under the direction of the HSSPF Eastlands and Russia-North (Jeckelen), had as its goal the elimination of the partisan bands in the Ossaveya Lake area. In a listing of his forces made by the HSSPF Eastlands and Russia-North on January 9, 1944, the 8th Police Armored Company is included as Police Armored Scout Car Company 8, seeing road-securing service in the Ossaveya Lake area.

At the end of July, in August and early September 1944 the 8th Police Armored Company served with the "Gieseke" (BdO. Eastlands, Generalmajor of Police Gieseke) Battle Group in the I. Army Corps, or with the "Held" Group (Hauptmann Held), subordinated to the 215th Infantry Division.

The strength of the company was reported on July 29, 1944 as two officers, 28 NCO and men, with four scout cars and one tank.

The "Gieseke" Battle Group reported in its daily orders of August 12, 1944 on its combat events:

"2.) Recognition
The Held Group, on August 11, 1944, has mopped up a Russian breakthrough with the strength of at least two battalions, despite constant very heavy enemy fire by artillery grenade launchers and antitank guns in a well-organized pincer counterattack, wiped out the mass of these two battalions, and reestablished

the old HKL. In the process, 28 prisoners (including one officer) were taken, 80 rifles, 30 machine pistols,5 heavy and 4 light machine guns, one 7.62 cannon, two 4.5 antitank guns, one radio set and much ammunition were captured and one 7.62 cannon destroyed.

I extend to Hauptmann Held for his outstanding personal bravery, the crews of the armored scout cars for their remarkable ability and cold-bloodedness, the commander of the Latvian Police Volunteer Battalion 23, Hauptmann Brigadiers, Hauptmann Fichtler, Intelligence Officer, Latvian Police Battalion 23, and all other officers, NCOs and men involved in this counterattack, my particular thanks and highest recognition. Through this brave deed the enemy's high-priority goal of breaking through the Memel front and thereby splitting the 215th Division was nullified."

According to the report on Ordnungspolizei forces and their actions, dated October 20, 1944, the 8th Police Armored Company was at that time subordinated to the BdO. Eastlands (Generalmajor of the Police Gieseke), at that time in Liebau and there the HSSPF Eastlands and Russia-North (SS Obergruppenführer General of the Police Jeckelen).

No other documents that mention the 8th Police Armored Company after October 1944 are known. According to a former member of the company, men of the company returned to Vienna-Purkersdorf before the war ended. Nothing is known of the fate of the armored vehicles.

Panhard armored cars are ready for action along a road. Since all the commanders are sitting in the turrets, their stop seems to be a brief one. The oval sheet-metal container at the rear served to hold supplies and equipment. (PFA)

Beside a Panhard armored car (Registration no. 271 66) of the 8th Police Armored Company, the platoon leader holds an action discussion. Except for the driver and a tank commander, all the men are wearing denim combination suits over their uniforms. (PFA)

A Panhard armored car, with fresh yellow delivery paint such as was used from 1943 on, has presumably just returned from servicing. Among the men in the foreground are members of a Schuma unit. (BAK)

In service with the company is this Russian BT-7 tank. The vehicle seems to have been overhauled carefully and has been painted yellow. Besides the German cross, the number 2 can be seen on the rear. A Panzer I tank was also in service with the company (see p. 242). (BAK)

9th Police Armored Company

Establishment, Structure and Equipment

The 9th Police Armored Company was established at the Police Motor Vehicle School in Vienna according to an order of the Reichsführer SS-Chief of the German Police dated November 18, 1942. The company was structured as follows:

Leader Group including supply train and workshop platoon

1st Armored Car Platoon (3 Panhard armored cars)

2nd Armored Car Platoon (3 Panhard armored cars)

3rd Tank Platoon (5 (Russian) tanks)

For the armored car platoons, Panhard vehicles that were already on hand at the Police Motor Vehicle School in Vienna were used. The tanks for the tank platoon were to be given to the company by the HSSPF Russia-Center. All other vehicles were assigned by special orders. The Police School for Technology and Transit Sub-armory in Vienna was to turn a (metal-wheeled) field kitchen over to the Police Motor Vehicle School in Vienna for the company. The assigning of the workshop platoon from the Police School for Technology and Traffic in Berlin was to be done during the course of the establishment time.

Assigning the motor vehicles to their home base was to be handled later, and the Ministry of the Interior registration numbers of the 9th Police Armored Company's vehicles were to be reported to the Reichsführer-Chief. The crews for the captured Russian tanks were to be instructed briefly on the tanks available at the Police School while the company was being set up. Their further training was to take place after the company had received the Russian tanks. All the forces intended for the company were to report to the Police Motor Vehicle School, Landstrasser Hauptstrasse 68, Vienna by 8:00 P.M. on December 9, 1942.

Before they were sent on the march from their German bases it was to be determined whether they were sufficiently healthy to stand the physical demands on tank crews. The inoculations and blood-group tests required by the orders of April 27, 1942 (MBliV. S. 804) and November 5, 1941 (MBliV. S. 1997) were to be checked and, if necessary, completed at once. They were to be supplied with clothing according to Appendix I of the order of July 12, 1941 (MBliV. S. 1303), amplified by that of August 5, 1942 (MBliV. S. 1644). The prescribed special clothing, namely a black coat, long black trousers, a black (fore-and-aft) field cap and a denim coverall suit, plus a Gas Mask 30 with breathing tube for each man, should be delivered by the P.V. in Vienna. Steel helmets would be used in place of crash helmets.

Each man was to be given a Pistol 08 or 7.65 mm with 80 or 50 bullets and an 84/98 bayonet by his home base. And the 98a carbines with 150 bullets each, needed for the further equipping of the Armored Company, were to be supplied by the Sub-armory in Vienna.

The members of the workshop platoon were to bring their rifles from their home bases. The establishment of the Armored Company was to be completed at the Police Motor Vehicle School in Vienna and reported to the Reichsführer-Chief by December 23, 1942. The order to set up the company was postponed briefly by the BdO. on December 10. The new starting date was December 12, 1942, and the process was to last until January 30, 1943.

According to orders from the Reichsführer-Chief in the Ministry of the Interior on January 4, 1943 (-O-Kdo. I K (2) 207 Nr. 231 III/43), Workshop Platoon 7, established at the Police School for Technology and Traffic in Berlin, with a strength of one master and seven policemen, was assigned to the 9th Police Armored Company and joined the company at the Motor Vehicle School in Vienna on January 29, 1943.

Action

According to a message from the Chief of the Ordnungspolizei on January 19, 1943 (KdO. I g I a (1) Nr. 58/43 (g)), the company left Vienna on January 31, 1943 for outside service with the HSSPF. Russia-Center. An advance command of one officer and five NCO had already been sent on the march on January 28, 1943. The company was quartered in Mogilev and subordinated to the SSuPolFhr. of Mogilev. Accordingly, it was also mentioned in several orders and reports as the Armored Scout Car Company, SSuPolFhr. Mogilev.

In the war structure report of the HSSPF. Russia-Center of April 20, 1943, the company was listed as "9. Pol.Pz.Sp.Kp." and its armored vehicles were erroneously listed as six Russian armored scout cars (which were surely French Panhards). In addition, the following captured Russian weapons were also on hand in the company: 12 light and ten heavy machine guns, one light grenade launcher, and two light antitank guns.

At this point there were still no armored vehicles with the company.

At the end of August 1943 Operation "Orloff" was planned in the zone of the 221st Security Division, since the northwest part of the division's area had been disturbed more and more threateningly by strong partisan bands for some time. Their daylight attacks on support points and vehicle columns had increased unbearably, to say nothing of the daily traffic disturbances and mine accidents. The strength of the bands was estimated at some 2000 bandits, their main forces constituting the "Grischin" band. The bandits were well and strictly led and well trained. The division received from the Commander of the Army Zone Center the order to surround and wipe out these groups, whose headquarters and masses, according to ground and aerial reconnaissance, were in the area north of Propoisk. To carry out this operation, the division was to use not only the troops in its own zone but also the Armored Scout Car Company, SSuPolFhr Mogilev.

The 9th Police Armored Company was assigned to the "Kopf" (Head) Group, consisting of the staff of Security Regiment 34, staff and second company of I./F.A.Grenade Regiment 636, East Battalions 642 and 643, 2./Security Regiment 930, I./East Company, Security Regiment 930, 1st Company, 286th Security Division, 15./East Company Security Regiment 45, Commander of the Flak Combat Troop with Light Flak Platoon, Enke Battery, 4th Artillery Group Boehme with staff and 2nd Battery, and East Battery 614.

The troops marched out to begin the operation on August 29, 1943, and the operation began on August 30. The "Kopf" Group took up their blockade line and, as planned, advanced with four armored scout cars of the 9th Police Armored Company as its spearhead, west of the Pronya through Rabovici and Uluki, took the crossing of the Resta and, by 7:00 A.M. reached, as ordered, the regiment's left boundary near Ushar. Troops of the 221st Security Division were able, on August 30, 1943, to surround the Grishin band completely, but the band broke out during the night of August 30-31. Attacks of the 221st Security Division on August 31 found nobody. Because of a transfer of the division, a planned continuation of Operation "Orloff" was called off.

On November 8, 1943 the 9th Police Armored Company was subordinated to Korück 559. Its initial strength was 101 men. its combat strength 44 men. Korück 559 planned an operation to free the Mogilev-Totoshin road early in October 1943, led by the 286th Security Division, and the 9th Police Armored Company was sent in between the 286th Security Division and SSuPolFhr. Mogilev of the 286th Infantry Division. The combat strength of the company was listed by Korück 559 on October 14, 1943 as 50 men.

On October 15, 1943 Security Regiment 45 of the 286th Infantry Division offered a suggestion for carrying out Operation "Entenjagd" (Duck Hunt, a second operation to mop up the Orsh-Toloshin-Belinichi-Mogilev area), in which the 9th Police Armored Company was to take part with three armored scout cars. In a combat report of Security Regiment 45 on October 16, 1943 on an Operation "Hasenjagd" (Rabbit Hunt) in the area north-northwest of Staroselye on the same day, the inclusion of the 9th Police Armored company was also reported.

On October 25, 1943 the "von Gottberg" Battle Group sent the order for Operation "Heinrich". Led by the chief of Partisan Combat Groups, an operation was carried out, its goal being the crushing of the "Partisan Republic of Rossonö in the Polozk-Krasnopolye-Pustoshka-Idriza-Sebesh area. Among the forces involved in Operation "Heinrich" were SS Police Regiment 2 (with its 1st Police Armored Company), SS Police Regiment 13 (with its 4th Police Armored Company), SS Police Regiment 24 strengthened by the 9th Police Armored Company, and the 12th (strengthened) Police Armored Company.

Reorganization as a Strengthened Police Armored Company.

Probably at the end of 1943 or the start of 1944, the company was restructured as a reinforced police armored company. Former members of the company report the addition of a platoon with Panzer I tanks. There is also mention of a motorcycle rifle platoon. Reorganization orders have not been found as yet. According to the available sources, the designation "9th Reinforced Police Armored Company (tmot.)" was used. The designations "9th Police Armored Company" and "9th Reinforced Police Armored company" were used interchangeably from then on in written messages. To date it has not been determined why the fully motorized unit was suffixed "partly motorized" (tmot). The structure of the strengthened company may have been as follows:

Leader Group including Intelligence Platoon, Supply Train and Workshop Platoon

1st Armored Car Platoon (3 Panhard armored cars)
2nd Armored Car Platoon (3 Panhard armored cars)
3rd Tank Platoon (5 Russian tanks)
4th Tank Platoon (5 Panzer I tanks)
Motorcycle Rifle Platoon

Action

On October 1, 1944 the 1st Army Corps took command over all troops and units located in the Polosk-Jasno Lake south end sector. Among them was the "von Gottberg" Battle Group with SS Police Regiments 2, 13, 24 and further police forces. In a structure report of the "von Gottberg" Battle Group on October 1, 1944, an armored scout car platoon of the 9th Police Armored Company is listed among Police Regiment 24.

The combat strength of this scout car platoon on January 10, 1944 was one officer and twenty men, the provision strength one officer and 27 men. In a structure report of the "von Gottberg" Battle Group as of February 29, 1944, the 9th Police Armored Company is listed along with the 1st and 12th Police Armored Companies. The "von Gottberg" Battle Group was subordinated to the 32nd Infantry Division at that point in time. On March 1, 1944, only one armored scout car platoon of the 9th Police Armored Company, with Panhard armored cars, was listed once again with SS Police Regiment 24.

The reinforced Police Armored Company 9 saw service in the area of the HSSPF. Russia-Center and White Ruthenia until mid-1944, but no precise data can be found.

The overview of Ordnungspolizei forces and their actions as of July 15, 1944lists only what remained of the company in the area of the HSSPF. Russia-Center and White Ruthenia, with its base in Schröttersburg, East Prussia. The strengthened Police Armored Company 9 (remnants) is listed under other units. The mass of the company was located in the zone of the BdO Danzig, stationed at Adlershorst. According to information from a former member of the company, it was quarantined there on account of typhus.

The overview of Ordnungspolizei forces and their actions as of October 20, 1944 still listed the strengthened Police Armored Company 9 as stationed in Adlershorst in the zone of the BdO Danzig under the HSSuPolFhr Vistula.

On October 26, 1944 the chief of the Ordnungspolizei (Kdo. g I Org/Ia (3) Nr. 769/44 (g.)) ordered a transfer of the 9th Strengthened Police Armored Company with all its forces, equipment, arms and vehicles out of the zone of the BdO, Danzig to the BdO Cracow, to be subordinated to SS Police Regiment 25. Transport was to be available as of October 30, 1944, arrival and quarters were to be reported by the BdO Kracow. In a report on the Ordnungspolizei in the Government-General as of November 17, 1944, the 9th Reinforced Police Armored Company was listed with its base in Yedrzeyov in the Radom district. The initial strength of the company at that time was four officers and 171 NCO and men.

On March 8, 1943 the 9th Police Armored Company was loaded onto the railroad and sent off to Mogilev, Russia. At left on the Panhard the vehicle's jack can be seen, on the car at right is an Stkw 14 truck. (ES)

The crew rode in freight cars and used every stop to open the big sliding doors and let air in, and to see something of the landscape. In March it was too cold to ride with open doors, and there were no windows. (ES)

A Panhard of the 9th Police Armored Company, with registration number 271 79, has chains mounted for travel in snow, but no winter camouflage. The Panhards of the 9th Police Armored Company had no front armor plates added at the bow. (ES)

Coming into a small village near Tula, the Panhard, with registration number 271 70 carries a prisoner on the engine cover. The crew of the scout car or the guards of the prisoner wear vehicle drivers' coats. (ES)

The crew is dressed in denim coveralls to service the vehicle. The radioman, wearing headphones, is working on the radio set. (ES)

Vehicles of the 9th (reinforced) Police Armored Company are seen in the yard of their quarters in Mogilev. In front is a T34 tank, behind it two British- or Canadian-built Valentine tanks. All vehicles have yellow camouflage paint with brown and green spots or stripes. (ES)

In the winter of 1943-44 the 9th (reinforced) Police Armored Company saw action in the Roslavl area. Wrapped in heavy winter uniforms, the crew of this Panhard poses for a photo. (ES)

It cannot be seen whether this T26 tank belongs to the 9th Police Armored Company or is an abandoned wreck. Since the Wehrmacht did not have enough towing capacity, not all the tanks shot down or burned out and abandoned were towed to the backland areas, even by 1943. (ES)

It is not clear to which unit this Panzer III, which broke through a bridge built by engineers, belonged, though the turret number suggests a Wehrmacht vehicle. (ES)

10th Police Armored Company

Establishment, Structure and Equipment

The establishment of the 10th Police Armored company took place on orders from the Reichsführer-Chief dated March 11, 1943 (O-Kdo. I K (2) 251 Nr. 71/43) at the Police Motor Vehicle School in Vienna. The company was to be structured like the 9th Police Armored company:

Leader Group including supply train and workshop platoon

1st Armored Car Platoon(3 Panhard armored cars)

2nd Armored Car Platoon (3 Panhard armored cars)

3rd Tank Platoon (5 (Russian) tanks)

On April 9, 1943, Armored Company 10 was assigned Workshop Platoon 27.

The armored vehicles to equip the 3rd Tank Platoon were lacking, and at first the crews were ordered back to their former bases. According to an order from the Reichsführer-Chief of December 2, 1943 (O-Kdo. In.K (1b) E.Nr. 72 II/43), the forces ready for the 3rd Tank Platoon of the 10th Police Armored Company were to march so that they would arrive at the Police Motor Vehicle School in Vienna by December 8, 1943. As a member of this platoon described it, the men were quartered in a small hotel in Vienna-Hadersdorf and trained on one captured Russian tank.

The formation and training of the 3rd Tank Platoon of the 10th Police Armored company was concluded on March 16, 1944. The 3rd Platoon was equipped with Russian tanks. Photos show Type T26 tanks.

Action

The 1st and 2nd Platoons and Workshop Platoon 27 were transferred to Russia shortly after their formation, and assigned to the HSSPF. Russia-South as the 10th Police Armored Company. The 10th Police Armored Company took part, as part of SS Police Regiment 10, in Operations "Vistula I" and "Vistula II" from May 12 to June 8, 1943, fighting against partisans in the "Wet Triangle" with the "Schimanä Battle Group.

After being subordinated to SS Police Regiment 11, the company took part in further anti-partisan operations and combat along with it. They were:

6/25-7/27/1943: Operation "Seydlitz" in the Olevsk-Perga-Borove-Begun-Slavechko area

8/1-9/23/1943: Operation "Shepetovkä

9/24-10/26/1943: Freeing the Slavuta-Shepetovka railway line

10/27-11/20/1943: Operation "Oslö

Early in 1944 the 10th Police Armored Company took part in front-line action. From January 1 to March 20, 1944 the company, along with SS Police Regiment 11, was with the XIII. Army Corps in defensive combat in the Kostopol-Rovno-Dubno-Kremianez-Brody area. The particular battles that qualified them for recognition with the Infantry Assault Medal (I) or Armored Combat Medal (P), are the following:

1/12/1944: Combat near Mala Lubaszka (I)

1/14/1944: Combat near Antonovka (I)

1/23/1944: Combat near Amelin (I,P)

1/26/1944: Counterattack near Rzeczyca (I,P)

1/28/1944: Counterattack near Hiazova (P)

1/29/1944: Counterattack near Berzoviec (P)

1/31/1944: Combat near Rzeczyca (P)

2/2/1944: Combat on the heights north of Rovno (P)

2/3/1944: Combat near Tynne ion the night of 2/3-4 (P)

2/5/1944: Combat on the Rovno-Dubno road near Dabrovka Colony (P)

3/4/1944: Assault on Hill 299, 3 km south of Sasnovka (P)

3/5/1944: Combat near Pisarovka (P)

3/6/1944: Assault on Hill 346.6 east of Kremianez (I,P); here a platoon leader's scout car was shot down; 3 crew members were wounded and rescued by another scout car's crew.

3/8/1944: Counterattack on Hill 330 north of Novosiolki (I,P)

3/15/1944: Combat near Czugale (P)

3/16/1944: Counterattack 2 km east of Lipovce (Kremianez) (I,P)

On March 17, 1944 the 3rd Tank Platoon, established and trained at the Police Motor Vehicle School in Vienna, was sent on the march to join the 10th Police Armored Company on orders from the Ordnungspolizei Headquarters. According to a former member of the 3rd Platoon, Shitomir was intended to be their destination, but because of the German retreat, the 3rd Platoon was underway by rail for weeks and finally tried to join the company in the Brody area via Lemberg. At first, though, this did not succeed.

Meanwhile, from April 22 to June 13, 1944 the majority of the company fought along with SS Police Regiment 11 and the 361st Infantry Division in defensive positions north of Brody. After that it was joined by its 3rd Tank Platoon in Lemberg, where it had seen no action until then. On July 7, 1944 the Chief of the Ordnungspolizei sent out an order (Kdo.g I Org./Ia (3) Nr. 179/44 (g)) to reorganize SS Police Regiment 11. According to this order, SS Police Regiment 11 was to be reorganized by the Commander of the Ordnungspolizei in Cracow. The police armored companies (7th and 10th Police Armored Companies) then with the regiment were to march to Vienna with all their vehicles, weapons and equipment for the purpose of refreshing their strength at the Police Motor Vehicle School.

In the overview of Ordnungspolizei forces and their actions as of July 15, 1944, the 10th Police Armored Company is listed in Wehrkreis XVII, the zone of the BdO. of Vienna. This report, though, applied as little to the 10th as to the 7th Police Armored Company (see also there), for the units were then with SS Police Regiment 11 to strengthen the defensive forces in the big bend of the Vistula. There the 10th Police Armored Company took part along with SS Police Regiment 11 in defensive fighting with the 174th and 214th Infantry Divisions. using both tanks and infantry. It took part in the following battles which counted toward recognition with the Infantry Assault Medal (I) and the Armored Combat Medal (P):

7/29/1944: Assault on Lucimia (P)
7/30/1944: Combat near Lucimia (P)
7/30/1944: Attack on Brzescie (P)
7/31/1944: Counterattack bear Brzescie (P)
7/31/1944: Combat on Hill 145 near Lucimia (P)
8/1/1944: Counterattack near Brzescie (P)
8/1/1944: Assault attack on Hill 145 near Lucimia (I,P)
8/2/1944: Counterattacks in Andrezeyov (I,P)
8/3/1944: Assault on Hill 149 west of Kucimia (I)
8/3/1944: Counterattack at the east end of Andrezeyov (I)
8/5/1944: Counterattack in Andrezeyov (I)
8/6/1944: Combat at the east end of Andrezeyov (I)

Former members report that the captured Russian tanks of the 3rd Platoon were mistakenly put out of action by their own air attacks in the Vistula bend. The platoon leader of the 3rd Platoon was lost in a scout-troop operation. No further data on the company's losses of men and materials in Russia are available. Like the 7th Police Armored Company, the 10th Police Armored Company was ordered back to Vienna for refreshing in September 1944.

Refreshing, Restructuring and Equipping

The overview of Ordnungspolizei forces and their actions as of October 20, 1944 lists the 10th Police Armored Company in Wehrkreis XVII, the zone of the BdO. in Vienna.

On December 4, 1944 the Chief of the Ordnungspolizei ordered (Kdo. g I Org/Ia (3) Nr. 892/44 (g)) the transfer of the 10th Police Armored Company from the Police Motor Vehicle School in Vienna to the BdO. of Italy. The 10th Police Armored Company, at full strength and with all its arms, equipment and vehicles, was sent by rail to the BdO. of Italy and subordinated to it. It was to be ready for transport as of December 7, 1944. The BdO. of Italy was to report the company's unloading depot, arrival and service location to the Chief of the Ordnungspolizei by December 6, 1944.

During its transfer to the Italian war zone, there were fighter-bomber attacks on the company.

According to former members of the company, the 10th Police Armored Company was unloaded in Lonigo (near Verona) and equipped there with Italian P40 tanks, some of which were broken down and had to be repaired by the workshop platoon. For the workshop platoon, this type of tank was completely strange.

Workshop Platoon 27 was assigned to the 10th Police Armored Company. At left is the low loader trailer, at right the repair shop truck and a sidecar motorcycle. At the right front is a Panhard with star antenna. (KG)

With the help of a block and tackle, parts of the motor are taken out of a Panhard. High demands were made on the workshop platoon, since getting replacement parts for the captured Panhards was not simple. (KG)

The 10th Police Armored Company presumably had a structure like that of the 15th Police Armored Company (see there):

Leader Group (2 P40 tanks)

1st Platoon (4 P40 tanks)

2nd Platoon (4 P40 tanks)

3rd Platoon (4 P40 tanks)

Repair Shop Group (Workshop platoon 27)

Replacement crews

Equipment and Baggage Train

A structure report of the HSSPF. Italy on April 9, 1945 gave the following data for the 10th Police Armored Company:

Subordination: SSPF. Upper Italy Center

Strength: 3/120

Armament; 15 tanks, 30 machine pistols, 50 rifles, 100 pistols.

Base: St. Michele

Action

During their withdrawal via Verona, Ponton, Rivalta, Ala and Rovereto, the 10th Police Armored Company used at least one tank platoon to cover the retreating moves of the Wehrmacht units.

A structure report of the HSSPF. Italy on May 6, 1945 gave the same data for the 10th Police Armored Company as that of April 9, except that the company was now based at Bozen.

According to former members, the company was taken prisoner by the Americans in Bozen and transported to Rimini. There the prisoners were turned over to British troops and taken to a prison camp in Grotalie near Tarento.

Panhard 271 87supports police riflemen advancing on a village occupied by partisans. For a better view, the commander observes from the opened turret hatch. (BAK)

Some houses in the village have been set afire. The tank commander wears the black officer's cap, which marks him as a platoon leader. He wears a motorcyclist's coat over his uniform for protection against the rain. (PFA)

Another picture of Panhard 271 87 of the 10th Police Armored company shows it advancing on the village. The police emblem can be seen clearly on the front of the turret. (BWS)

Panhard 271 87 crosses a bridge or causeway made of tree trunks. A second Panhard, with number 271 84, follows it.

Police tanks are seen crossing a ditch. Panhard 271 84 has V-shaped armor added in front, plus an extra headlight mounted behind it. (BAK)

Three T26 tanks of the 10th Police Armored Company in 1944, with 1939 models at left and right and a 1933 model T26 in the middle, with the older turret and structure but the same armament. (KW)

Men of the 10th Police Armored Company pose in the Bozen-Verona area at the end of 1944. It appears that they are wearing tropical uniforms. (KW)

A P40 tank, 15 of which the 10th Police Armored Company received in May 1945. The vehicle shown here was not used by the company, but tested at the Army Weapons Office. (BAK)

11th Police Armored Company

Establishment, Structure and Equipment

The 11th Police Armored Company was established by order of the Reichsführer SS and Chief of the German Police in the Ministry of the Interior on August 23, 1943 (O-Kdo. I K (1b) 251E Nr. 21/43). Accordingly, the company was structured as follows:

Leader Group with supply train and Workshop Platoon 29
1st Platoon (3 Panhard armored cars)
2nd Platoon (3 Panhard armored cars)
3rd Platoon (5 Hotchkiss tanks)

It has not yet come to light why the 11th Police Armored Company was established so much later than the 10th and 12th Police Armored Companies. On September 19, 1943, men already trained on Panhard armored cars by the Armored Replacement Unit were ordered to join the 11th Police Armored Company at the Police Motor Vehicle School in Vienna; and the equipping and training of the company began at this point. According to a command of the Reichsführer-Chief of September 25, 1943 (O-Kdo. I K (1b) 251 E Nr. 16II/43), Workshop Platoon 29, formed at the Police School for Technology in Berlin, with one master and seven NCO and men, was to join the company in Vienna. By order of the Chief of the Ordnungspolizei on November 20, 1943 (Kdo. gI-Ia (1) Nr. 892III/43 (g)), the 11th Police Armored Company was to be sent and subordinated to the BdO. Croatia. The march from the Police Motor Vehicle School in Vienna to Croatia (KdO. E 1550/43 XL) began on November 30, 1943.

On June 3, 1944 the Chief of the Ordnungspolizei ordered (Kdo. g I Org. (3) Nr. 109/44 (g)) the formation of engineer units at a strength of 1/10 for the 6th and 11th Police Armored Companies (see also 6th Company). The Police Weapons School I (PSW I) in Dresden-Hellerau reported on June 14, 1944 that the engineer units' training has been completed, and on orders from the Chief of the Ordnungspolizei (Kdo. g I-Ia (1) Nr. 176/44 (g)) on June 19, 1944, they were sent on the march to Croatia.

Action

As of December 1943 the Company was used against partisan bands in the zone of the BdO. Croatia, Police Region V,k usually subordinated to SS Police Regiment "Croatia".

From March 12 to 17, 1944 the Company took part in Operation "Falke" (Falcon), fighting against partisans in the Valpovo-Djakovo area, the Bila Gora, the Papuk Mountains and Sedlarica-Pitomaca. Operation "Falke" was carried out since February 20, 1944 in the zone of the LXIX. Army Corps z.b.V. as a combing operation. Participants were SS Police Regiment "Croatiä, Police Battalions 6 and 8, Police Armored Company 11, and Police Gun Battery 2, plus the forces occupying Vitrovitica (Police Battalion 9) and the German Gendarmerie support points in the area.

From April 6 to 25, 1944, the company participated in Operation "Osterhase" (Easter Bunny), racapturing Podr. Slatina and fighting partisans in the central Drava area of Croatia.

From April 26 to May 20, 1944 it participated in Operation "Ungewitter" (Storm), fighting partisans in the Bila Gora and Papul Mountains. Operation "Ungewitter" was carried out by the 42nd Jaeger Division, along with the 4th Croatian Mountain Brigade, Ustascha units, and a battle group of Police Regiment 1, commanded by the staff of the RFSS for Croatia, with Police Battalions 6 and 9, 5./ Engineer Battalion 86, 11th Police Armored Company, 2nd Police Gun Battery, Gendarme Platoons (mot.) 34 and 67, and Ustacha units. The company also took part in Operation "Kornblume I" (Cornflower), fighting partisans in the Sid-Sidski-Banovci-Lipovac-Marovic area. From June 12 to 22, the company took part in Operation "Kornblume II", fighting partisans in the Fruska Gora.

The overview of Ordnungspolizei forces and their actions as of July 15, 1944 lists the 11th Police Armored Company with its minesweeping troop under the HSSPF. Croatia in the zone of the BdO. Esseg, stationed in Agram.

The overview of October 20, 1944 lists the same situation, with a base in Nasice. At this time, the 6th and 16th Police Armored Companies (see there) were in the zone of the BdO. Esseg along with the 11th Company, each with a minesweeping troop.

In December 1944 the 11th Police Armored Company was on frontline duty with the XXXIV. Army Corps or the Kübler Corps Group in the area of the Police Security Zone South. According to a message from the Police Security Zone South to the Kübler Corps Group, the 8th Croatian Jäger Brigade, parts of the I./ Police Regiment 5, the 1st Police Gun Battery and the 11th Police Armored Company were subordinated to the Kübler Corps Group for an attack on Ivanovci, Tamasanci, Gorjani and Satnica on December 14, 1944. The troop structure list of the Police Security Zone South showed for the same period the 6th Police Armored Company and Police Armored Platoon 1 (see there) as subordinated armored forces

On December 19, 1944 the Division z.b.V. Stephan of General Command, XXXIV. Army Corps, listed Djakovo NO (Tomislava Street) as the base of the 11th Police Armored company, which was subordinate to it.

The 11th Police Armored Company remained subordinate to the Police Security Zone South in 1945, and is referred to under various code names (such as Leierkasten, Katzenjammer, Männerriege) in the intelligence officer's radio records from January 1 to March 21, 1945. The 11th Police Armored Company was last mentioned in the records of the Intelligence Leader, BdO. Croatia, on April 20, 1945 (valid to May 1, 1945).

Nothing is known about its end.

A Panhard armored car and a Hotchkiss tank of the 11th Police Armored Company are being transported by rail to Croatia in the autumn of 1943. The tank's star antenna is easy to see. (AF)

The crew of a Hotchkiss stand by their tank, which is loaded on a flatcar in the winter of 1943-44. All Hotchkiss tanks used by the Ordnungspolizei had a commander's cupola with a hatch cover. This modification of the original was carried out by the Wehrmacht before the vehicles were turned over to the Ordnungspolizei. (ML)

Two Hotchkiss tanks are shown in Croatia in the winter of 1943-44. The vehicles are not yet camouflaged for the winter. The tank in front had its engine cover raised, which suggests work on technical problems. (ML)

Members of the 11th Police Armored Company are having a snowball fight in front of their quarters in Croatia. (AF)

12th Reinforced Police Armored Company

Establishment, Structure and Equipment

According to reports from former members of the company, it was probably established in October 1942 at the Police Motor Vehicle School in Vienna. According to records of the HSSPF Russia-Center, the company was already listed there as a "reinforced" company in January 1943, so that it can be assumed that the company was first established as the 12th (reinforced) Police Armored Company. Only reports from former members and photos tell of its equipment, indicating that the company was structured as follows:

Leader Group with Supply Train and Workshop Platoon
1st Platoon (3 Russian armored cars)
2nd Platoon (3 Russian armored cars)
3rd Platoon (5 Russian and Panzer III tanks)
4th Platoon (5 Russian tanks)

The armored vehicles were not received until the company reached its base at Mogilev on the Dniepr, and had to be repaired by its own workshop platoon. Among the vehicles used, according to reports, there was a Russian T 34 tank. Photos also prove the existence of Russian BA 10 scout cars, German Panzer III tanks, Russian T 60 and T 70 tanks, and one T 26.

Action

As of December 22, 1942 the company saw action outside Germany and was subordinated to the HSSPF. Russia-Center. The company's first real base was Mogilev; later in 1943 it was transferred to Smolevice and saw action at various places in the central sector of the eastern front. From March 30 to April 7, 1943 the 12th Police Armored Company was subordinated to the SS Dirlewanger Battalion for Operation "Lenz-Sud" (Spring-South), fighting against partisans in the Borisov-Sloboda-Smolevice area. Other forces involved included Police Regiment 13, I./Police Regiment 23, Patrolman Battalions 57 and 202.

For Operation "Draufgänger II" (Daredevil) in the Manily and Rudina Forests from May 1 to 10, 1943, Police Armored Company 12 was assigned to Police Regiment 2 and subordinated to the 1st (reinforced) Police Armored Company.

The overview of forces by the HSSPF. Russia-Center and White Ruthenia on April 1, 1943 listed the 12th (reinforced) Police Armored Company as being subordinate to the SSPF. White Ruthenia for service and tactically to the "Schimanä Battle Group. Its base was given as "in service". The next such list, of May 1, 1943, listed its service and tactical subordination to the SSPF. White Ruthenia; its base was still stated as "in action". In July 1943 the 12th (reinforced) Police Armored Company participated in Operation "Hermann" with the "von Gottberg" Battle Group. Operation "Hermann" took part in the Sluzk-Baranovice-Novogrodek area. The battle group's command post was in Novogrodek on July 15, 1943. Troops involved included the 1st SS Infantry Brigade, SS Police Regiment 2, Gendarme Platoon (mot) 21, and the 12th (reinforced) Police Armored company, among others.

The company remained subordinated to the "von Gottberg" Battle Group and took part with it, in August, in the evacuation of the Jeremicze-Starzyna-Krezrczoty-Rudnia-Czartovice-Potasznia-Delatycze-Kupisk area. From August 11, 1943 on, the company was assigned to the Novogrodek Area Commissar for seizing purposes. Fort Operation "Zauberflöte" (Magic Flute) the 12th Strengthened Police Armored Company was subordinated to SS Police Regiment 13.

In an order dated August 14, 1943, the HSSPF. Eastlands and Russia-North awarded four Iron Crosses second class for special bravery in the face of the enemy to members of the 12th (reinforced) Police Armored Company. The company or parts of it had thus seen service in the Russia-North area at that time or previously.

On October 25, 1943 the "von Gottberg" Battle Group gave the order to begin Operation "Heinrich". In this operation, led by the chief of partisan band fighting, an action was to be carried out to crush the "Rossono Partisan Republic" in the Polozk-Krasnopolye-Pustochka-Idriza-Sebesh area. Among the forces taking part in Operation "Heinrich" were SS Police Regiment 2 (with 1st Police Armored Company), SS Police Regiment 13 (with 4th Police Armored Company), SS Police Regiment 24 reinforced by the 9th Police Armored company, and the reinforced 12th Police Armored Company.

On January 10, 1944 the First Army Corps took command of all the troops and units in the Polozk-Jasno Lake South End area to form a front line against the Russian Army. [p. 130] Among these units was also the "von Gottberg" Battle Group, with the police forces listed above for Operation "Heinrich". In the war structure report of the "von Gottberg" Battle Group of January 10, 1944, the fighting strength of the 12th Police Armored Company was listed as one officer and 29 men, and its total strength as one officer and 43 men. Probably this was just one platoon of the company, being listed in the schematic structure list as having one Armored Scout Car (wheeled), one armored scout car (tracked) and one tank.

On March 1, 1944 the 12th Police Armored Company was still listed along with the 1st and 9th Police Armored Companies in the war structure report of the "von Gottberg" Battle Group, which was then subordinate to the 32nd Infantry Division of the Army. The 12th Police Armored Company, or its tank platoon, formed the battle group's reserve in its structure reports of January and March.

According to reports of former members of the company, after front service near Vitebsk in the winter of 1943-44 it returned to its base at Smolevice. Various documents show that the 12th Police Armored Company was under the HSSPF. Russia-Center in the spring and summer of 1944. The overview of Ordnungspolizei forces and their actions as of July 15, 1944 listed the company as an independent unit in the area of the HSSPF. Russia-Center and White Ruthenia, with its base then in Schrötersburg, East Prussia.

Former members of the company report that it passed through Vilna and into East Prussia during the retreat in the central sector of the eastern front. From there it was sent to Frankfurt on the Oder, where repairs were made to its still-extant vehicles. In late July and early August 1944, the company was sent to Pressbaum, near Vienna-Purkersdorf, to the Police Motor Vehicle School-Police Armored Replacement Unit, arriving there on August 13, 1944.

Reorganization and Reequipping as an Armored Company

The 12th (reinforced) Police Armored Company was reorganized at the Police Motor Vehicle School-Police Armored Replacement Unit in Vienna beginning in September 1944 (Kdo. In K Ib 251 E Nr. 197/44 of 8/31/44 and Kdo. In K Ib 251 E Nr. 206/44 of 9/15/44). Parts of the company were sent to the Armored Replacement Unit South in Lonigo near Verona in Italy for training on Italian tanks. The BdO. Verona still listed the company in a message dated November 9, 1944 as having 74 men in the units provisioned there. The restructuring of the 12th (strengthened) Police Armored Company took place according to War Strength Directive (Army) Nr. 1149, Section B "Assault Gun Unit with 14 guns" (as appendix). Not listed were the forces called for in the war strength directive for the recovery troop and the Division Workshop Company (mot).

But the company was not armed only with assault guns. Photos show that it also had Italian M 15 tanks. It is possible that the 12th (reinforced) Police Armored Company took over the materials that were originally intended for the 15th Police Armored company. The latter was a totally new unit, while the 12th Police Armored Company already had combat experience. Thus a unit ready for action could be created more quickly.

After the restructuring, the 12th (reinforced) Police Armored Company was designated 12th Police Armored Company and then consisted of:
Leader Group (1 M 15 tank, 1 M 42 gun)
1st Platoon (4 M 15 tanks)
2nd Platoon (4 M 42 guns)
3rd Platoon (4 M 42 guns)
Repair Troop (Workshop Platoon 20)
Replacement crews
Combat and Supply Train
Its strength in personnel consisted of four officers and 111 NCOs and men.

Action

As early as September 27n 1944 the Chief of the Ordnungspolizei ordered, with great urgency (Kdo. g I Org/Ia (3) Nr. 628/44 (g.)), the transfer of the 12th Police Armored Company from the Police Motor Vehicle School-Police Armored Replacement Unit in Vienna to the BdO. Belgrade and its subordination to SS Police Regiment 5. Readiness for transport was to be reported at once. But this transfer did not take place, for on

October 21 the Chief of the Ordnungspolizei canceled this order (Kdo. g I Org/Ia (3) Nr. 628II/44 (g.)) and ordered the company, in full strength, with arms and equipment, sent to the HSSPF. Hungary and subordinated to it as soon as possible. Transport readiness was to be reported by October 25, 1944. On October 27, 1944 the 12th Police Armored Regiment left the Police Motor Vehicle School in Vienna for outside service with the BdO. Budapest. The overview of Ordnungspolizei forces and their actions as of October 20, 1944 anticipated reality somewhat and listed the 12th (strengthened) Police Armored Company (still under its old designation) as being with SS Police Regiment 1 in the zone of the BdO. Budapest under the HSSPF. Hungary.

In combat in Budapest and the surrounding area, the 12th Police Armored Company was wiped out by about the beginning of 1945.

On March 27, 1945 the Chief of the Ordnungspolizei ordered (Kdo. In K 1b 251 Nr. 147/45) the disbanding of the 12th Police Armored Company. The company's surviving personnel were sent back to the Police Motor Vehicle School-Police Armored Replacement Unit and made available as replacements.

All the remaining vehicles and equipment were taken over by the Police Motor Vehicle School in Vienna. Members of Workshop Platoon 20, with their remaining vehicles and equipment, were to march to the Police Motor Vehicle School-Workshop Replacement Unit in Iglau.

A heavy Russian armored car and three light T60 tanks of the strengthened 12th Police Armored Company are seen in winter camouflage. The armored car (Wehrmacht designation: Armored Scout Car BA 203 (r)) is marked only with a small German cross. The column's unarmored vehicles have no winter camouflage paint. (MG)

Two of the company's T60 tanks pass a burning village. The small tanks have the company emblem painted on the right front armor, with German crosses on the bow plate and the big turret hatch. (EH)

This Stkw. 4 (Pol-64455) is seen crossing a ford in Russia. At the left rear, two of the company's armored vehicles can just be seen. A heavy Russian 6-wheel armored scout car is followed by a twin-turret T26 tank, Model 31. (EH)

Soldiers rest along a marshy road while a horsedrawn battery of 7.62 mm infantry guns passes by. On the saddlebag of the solo cycle the emblem of the 12th (strengthened) Police Armored Company can be seen. (EH)

Until mid-1943 command radio trucks were rare among the armored police companies. The company chiefs of individual units, such as the 12th Police Armored company seen here, could keep in touch with their platoons via portable radio sets. (EH)

"Riding lesson" on a very skinny horse, with two heavy Russian armored cars behind the rider, followed by a T60 tank. The front armored car has been camouflaged with branches. (MG)

A Panzer III of the 12th Strengthened Police Armored Company is in a column behind a T60 tank. The Panzer III Type G is armed with a 3.7 cm gun and two coaxial machine guns in the turret. Another machine gun mounted at the right front was operated by the radioman. (MG)

Another Panzer III of the company, this time Type G armed with a 5 cm gun and one coaxial machine gun in the turret, plus a machine gun in the mount ahead of the radioman. The Panzer III is followed by a T7- tank, armed with a 4.5 cm gun and a machine gun in the turret. (MG)

A Panzer III Type G and two Russian heavy armored cars; the latter, armed with a 4.5 cm gun and a machine gun in the turret, plus a machine gun at the right front, were more heavily armed than the Steyr and Panhard armored cars. (MG)

In the autumn of 1944 the 12th Strengthened Police Armored Company became the 12th Police Armored Company and was equipped with Italian M15 tanks and M42 assault guns. Typical of the ex-Italian M15 tanks are the riveted plates and the twin bow machine guns.

In December 1944 this M15 tank of the 12th Police Armored Company was advancing on Russian positions in the Budapest area, followed by Hungarian infantry.

This M42 assault gun, seen at a scrap depot east of Budapest in March 1945, probably belonged to the 12th Police Armored Company, which had nine of these vehicles, each armed with a 7.5 cm gun. (MB)

13th Reinforced Police Armored Company

Establishment, Structure and Equipment

The 13th Police Armored Company was organized as a strengthened company on the basis of an order from the Reichsführer SS and Chief of the German Police in the ministry of the Interior (O-Kdo. I K (2) 251 Nr. 3/43) on January 6, 1943. It was not established as usual at the Police Motor Vehicle School in Vienna, but with the "Griese" SS Police Regiment on the Atlantic coast of France.

By Order Nr. 3267 of the Reichsführer-Chief on February 15, 1943 (O-Kdo. I K (2) 205 Nr. 10/43) it was ordered that two Steyr armored scout car platoons organized for the 7th Police Armored Company, with all their personnel and vehicles, were assigned to the 13th (reinforced) Police Armored Company with the SS Griese Police Regiment. No sources assigning the other vehicles have been found. The establishment and training of the company took place at Marseilles from mid-February on. The company's structure was as follows:

Leader Group with Supply Train and Workshop Platoon
1st Platoon (3 Steyr armored cars)
2nd Platoon (3 Steyr armored cars)
3rd Platoon (6 Panzer II VK 1601 tanks)
4th Platoon (4 Panzer IV Type F1 tanks)
Flame throwing Armored Car Platoon (2 Sd.Kfz. 251/16)

The company's first base was the "Les Beaumettes" villa colony near Marseilles, where the approximately 200 men (original strength as of 6/22/1943: 188 men) were housed in eleven small summerhouses.

Steyr 227 74 with its motorcycle courier are seen during off-road training in France in 1943. The German cross for air recognition can be seen atop the turret. At left on the bow is the troop emblem of the 13th reinforced Police Armored Company. (BAK)

Action

The company was in service with the "Griese" Police Regiment on the south coast of France until July 11, 1943, taking part in, among others, the evacuation and destruction of the Marseilles harbor area.

Along with the "Griese" Police Regiment, then renamed SS Police Regiment 14, the unit was transferred to Croatia in July 1943.

From July 14 to August 5, 1943 the unit was active in anti-partisan activities in southern Croatia, in areas including the Fruska-Gora. Afterward it saw anti-partisan service in northern Croatia, in the Bila-Bora and other localities. As of September 10, 1943 it advanced to the Adriatic coast to disarm Italian troops. There followed defensive combat around Ogulin and Ostarje. From September 28 to October 10 the company was surrounded by strong enemy forces at Ostarje.

From October 20 to November 13, 1943 it took part in the pacification of Slovenia, taking part in Operation "Wolkenbruch" (Cloudburst) and serving under the "Hoch-und Deutschmeister" Reichsgrenadier Division. On November 14 and 15, 1943 it fought to relieve Rudolfswerth, and then fought in the Rudolfswerth area until November 28, 1943.

As of November 29, 1943 it took part in the pacification of the Oberkrain, lower Styria and the Laiback province. Here the company served with SS Police Regiment 14 and at times also with SS Police Regiment 19 under the HSSPF. in Defensive Zone XVIII "Command Staff for Partisan Combat", and was stationed in Laibach.

The overview of Ordnungspolizei forces and their actions as of October 20, 1944 still lists the strengthened Police Armored Company 13 with SS Police Regiment 14 in the zone of the BdO. Salzburg, HSSPF. Alpine Lands. At the same time, SS Police Regiment 14 also served there with the 4th Police Armored Company and the 14th (reinforced) Police Armored Company. Bases and subordinations of the 13th (reinforced) Police Armored Company after October 1944 are not known. In 1945, SS Police Regiment 14 was used to organize the 35th SS Police Division, but whether the 13th (reinforced) Police Armored company was still subordinate to the regiment at that time is not known.

All six Steyr armored cars of the 13th (reinforced) Police Armored Company are seen driving past a small harbor. The vehicles had originally been intended for the 7th Police Armored Company and were all from the first series, taken over from the Austrian Army or Police. (BAK)

An armored car platoon of the 13th Strengthened Police Armored Company is seen on the coast near Marseilles in the summer of 1943. The vehicles did not have either bow machine guns or radio antennas, installed later, mounted at that time. For lack of radios, each armored car has a motorcycle courier assigned. (BAK)

The company's armored cars and tanks stop in a town near Marseilles. The armored car in front still has an emblem on the turret, which was no longer common at that time. Some tools can be seen attached to the bow armor. (BAK)

The company's Panzer IV tanks were fairly new and well armed with short 7.5 cm tank guns. In the Wehrmacht's tank units, though, the long-barreled Panzer IV had been introduced already. The Panzer IV and T 34 tanks were then the police's heaviest weapons. (BAK)

The company's Panzer II and IV tanks are seen from the rear. They are all painted gray and bear German crosses, registration numbers and unit emblems. At the front is a Panzer II (VK 1601), commanded by the platoon leader, an Oberleutnant of the Schutzpolizei, with an honor chevron on his uniform. (BAK)

Steyr armored cars pass a turret emplacement guarding a concrete tank wall in the background. The wall guards the entrance to the town from attempted tank landings on the coast. The turret came from a Hotchkiss or Renault whose body was used by the Wehrmacht as a tractor or tank destroyer. (BAK)

Another photo of the company's passage through a town. The Panzer II tanks are following the company's flame throwing tanks. In front of the three boys in shorts is the entrance to the turret emplacement's bunker. (BAK)

Two Panzer II (VK 1601) of the 13th Strengthened Police Armored Company. Six of these heavily armored vehicles, of which only 30 were built, saw service with the company. (BAK)

The two flame throwing tanks (Sd.Kfz. 251/16) return to the company's base near Marseilles. On each side of the bodies are the big, fixed flame tubes, and at the back is the small mobile flame tube, that could be used up to the hose's length away from the vehicle (BAK)

14th Police Armored Company

Establishment, Structure and Equipment

The 14th Police Armored Company was organized according to an order from the Reichsführer SS and Chief of the German Police in the Ministry of the Interior on July 5, 1943 (O-Kdo. I K (Ib) 251 Nr. 279/43). The date and place of organization are not known, but the process was finished by the end of October. Another order, dated April 5, 1944 (O-Kdo. In K (Ib) 251E Nr. 125/44), probably refers to the assignation of Italian armored scout cars and assault guns and the training of crews, but the contents of the order are not known. At first the company was probably structured as follows:

Leader Group with Supply Train and Workshop Platoon 28
1st Platoon (3 Tatra armored cars)
2nd Platoon (3 armored cars from Holland)
Rail Armored Platoon (4 light armored rail cars)

To form the 1st Platoon, the Armored Scout Car Platoon of the Police Action Staff Southeast in Laibach (see there) and its three Tatra armored cars were used. The origin and type of the armored scout cars from Holland are not known, nor is the origin of the armored rail cars.

Action

At the end of October 1943 the company was subordinated to the Commander of the Ordnungspolizei-Command Post Veldes and stationed in Laibach. The 14th Police Armored Company was directly subordinated to the HSSPF. of Defensive Zone XVIII "Command Staff for Partisan Combat". It spent all of 1944 in the zone of the BdO. Salzburg (Defensive Zone XVIII). In this area, SS Police Regiment 14 and the 13th (strengthened) Police Armored Company were in service at the same time, with SS Police Regiment 13 and the 4th Police Armored Company in southern Carinthia.

Restructuring as a Reinforced Police Armored Company

On September 19, 1944 the Chief of the Ordnungspolizei gave an order for the reorganization of the 14th Police Armored Company (O-Kdo. In K (Ib) 251E Nr. 125II/44). According to the accompanying strength and equipment instructions, the company was thus to be turned into a strengthened police armored company. All the armored vehicles were already with the company at that time. Newly assigned were one Stkw. 8 (without radio equipment) as a command radio car, one Stkw. 14 each for the 4th and 5th platoons, and two trailers to carry baggage. Nine supply trucks, which could not be provided, were to be taken from available captured vehicles supplies. These could well have been Italian trucks.

The lacking personnel were to be supplied by the Police Motor Vehicle School-Police Armored Replacement Unit in Vienna: one armored platoon leader (assault guns), six tank commanders and simultaneously radiomen (for assault guns), one intelligence platoon leader, one intelligence device operator, one radioman, four tank gunners, one driver for a Tatra armored car, 12 drivers and one motorcycle rifleman. Also three locksmiths for Workshop Platoon 28. All forces were to be sent on the march to the company on October 5, 1944.

The 14th (reinforced) Police Armored Company was, according to the structure report of September 19, 1944, the police's strongest armored company, with:

Leader Group with Intelligence Platoon, Supply Train and Workshop Platoon 28
1st Platoon (3 Tatra armored cars)
2nd Platoon (3 AB 41 armored cars)
3rd Platoon (3 armored cars from Holland)
4th Platoon (8 L6 assault guns)
5th Platoon (8 L6 assault guns)
Reserve Half-platoon (4 L6 assault guns)
Rail Armored Car Platoon (4 light armored rail cars)

According to the specified strength, there were also two Stkw. 4, two Stkw. 8 (one of them a radio command car), five Stkw. 14, one workshop truck, one loading aggregate, one towing truck with low-loader trailer, one field kitchen, 22 Wkw. (two with trailers), and 16 motorcycles (8 with sidecars) for the company.

The specified manpower amounted to six officers, one administrative official, 43 NCO, 80 men and 119 motorcyclists. Training on the Italian armored vehicles took place, according to several former members, in Verona with the Wehrmacht (probably at the Armored Training Unit in Lonigo).

Action

The sphere of action for the 14th (reinforced) Police Armored Company remained the Laibach province or the command area of the BdO. Salzburg.

In 1945 the 14th (reinforced) Police Armored Company saw service at the front. This was reported by the Commanding General of Backline Area E on April 28, 1945: SS Police Regiment 17, SS Police Regiment "Todt" 28 (1st & 3rd Battalions), SS Gendarme Battalion 3 (mot), and the 1st and 14th Police Armored Companies were in action on the Goettenitz-Gottschee Line as far as Gurk (Topler Reber).

According to information from former members, the 14th (reinforced) Police Armored Company was stationed at the Drau bridge when the war ended.

These two Italian armored vehicles of the 14th (reinforced) Police Armored Company were photographed in Laibach Province in 1944. The AB 41 also went into series production with turret headlights and heavier front armor. (BP)

The company's L6 Italian assault guns were armed with a 4.7 cm gun as their primary weapon, plus a machine gun behind a shield on the top of the vehicle. This camouflage paint is typical of Italy and the Balkans. (BP)

144

In the winter of 1944-45, this Lince armored car saw service with the 14th (reinforced) Police Armored Company. The small vehicle, copied from the British Daimler Scout Car, had a two-man crew and was armed with a machine gun. (BP)

The light armored rail car was seldom seen with the Ordnungspolizei, though the 14th (reinforced) Police Armored Company had a platoon with four of them. (BP)

15th Police Armored Company

Establishment, Structure and Equipment

The 15th Police Armored Company was founded on the basis of an order from the Chief of the Ordnungspolizei (O-Kdo.g In K (1b) Nr. 46/44 (NfD)) of July 11, 1944. The chief of the Ordnungspolizei referred to a command of December 12, 1943 (O-Kdo. In K (1b) 205 E Nr. 21 III/43) and a message of July 1, 1944 (O-Kdo. In K (1b) 251 E Nr. 157/44). The first Company Chief of the 15th Police Armored Company entered service with the company on April 7, 1944.

According to the order of July 11, 1944, the 15th Police Armored Company was to be organized under the Commander of the Ordnungspolizei in Italy according to the Army's War Strength Directive No. 1149, Version B, with 14 armored vehicles and captured equipment assigned to the Formation Staff South. But the War Strength Directive applied only as a starting point, and the towing troop of the repair group and the manpower for the division workshop company were eliminated.

A motorcycle rifle platoon was to join the company according to Army War Strength Directive No. 1112, Section 1 (1st Platoon).

Workshop Platoon 43, already with the company, was to serve as the motor vehicle maintenance group. This shows that at that time the company was already partly assembled and presumably occupied with repairing captured Italian equipment.

By order, the company was made up as follows:

Leader Group (1 M 15 tank, 1 M 42 assault gun)
1st Platoon (4 M 15 tanks)
2nd Platoon (4 M 42 assault guns)
3rd Platoon (4 M 42 assault guns)
Maintenance Group (Workshop Platoon 43)
Exchange Crew
Combat and Baggage Train
Motorcycle Rifle Platoon

According to Directive 1149 Version B, there were seven Stkw. 4, one Pkw. 4 car, four sidecar motorcycles, four two-ton, ten three-ton and two 4.5-ton trucks, and a one-ton towing tractor.

The full personnel strength consisted of three officers, 41 NCOs and 69 men. For the motorcycle rifle platoon there were, according to Directive 1112 Section B, two Stkw. 4, one cycle, 12 sidecar cycles and one truck, plus one officer, five NCOs and 38 men. Of the needed vehicles, the Police Motor Vehicle School of Vienna was to provide two Stkw. 4 and four motorcycles. The trucks were to come from the BdO. Italy from those trucks turned over by the OKH. A field kitchen, when needed, was to be requested from the Chief of the Ordnungspolizei.

The officers noted in the command of July 1, 1944 (O-Kdo. In K(1b) 251 E Nr. 157/44) were to serve as platoon leaders. Fifteen NCOs and men were to come from the Police Motor Vehicle School in Iglau to serve as drivers, plus four as motorcycle drivers. One NCO medic was to come from the Police Medical Training and Replacement Unit in Tabor, and a shoemaker and a tailor from the PV in Berlin.

These NCO and men, who could not be born earlier than 1901, were to report to the Police Motor Vehicle School-Armored Replacement Unit in Vienna on July 25, 1944 to take charge of the vehicles and take part in transport. But if there were still openings in the company despite the assigned forces, these could be filled, if urgently needed, by request to the Chief of the Ordnungspolizei.

The establishment of the motorcycle rifle platoon was to take place with available personnel at the Police Motor Vehicle School-Armored Replacement Unit in Vienna. An engineer NCO was added to the troops of the motorcycle platoon, and 15 motorcycle riflemen were to be trained as engineers. The vehicles needed for the motorcycle platoon were to be made available from supplies at the Police Motor Vehicle School in Vienna. Likewise, a set of engineer tools was to be supplied to the motorcycle platoon. Instead of the six light machine guns called for in the strength directive, the platoon received only three light machine guns.

The Police Motor Vehicle School (Armored Replacement Unit) was also to supply the police armored car crews with service clothing. The NCO and men were to be supplied with a Pistol 08 or 7.65 mm and 80 or 50 bullets each, an 84/98 bayonet and a Gas Mask 30. The further equipping of the company with weapons called for in the war strength directive was to be done by the BdO. Italy.

All forces and motor vehicles at the Police Motor Vehicle School in Vienna were to be ready to be called by the BdO. Italy as of August 23, 1944.

On September 13, 1944 the Chief of the Ordnungspolizei ordered (Kdo. In k (1b) 205 E Nr. 64/44), referring to the order of July 11, 1944 to the Police Motor Vehicle School-Armored Replacement Unit in Vienna, the sending of a maintenance technical sergeant, a camp administrator, a fuel attendant, a radio repairman, three observation NCOs and one tank driver for the 15th Police Armored Company.

Whether the 15th Police Armored Company ever received the armored vehicles listed in the July 11, 1944 order according to plan cannot be proved at this time. Presumably the M 15 tanks and M 42 assault guns of the 12th Police Armored Company were taken over in September 1944.

The 15th Police Armored Company was sent P 40 tanks as its new supply of tanks. They probably could not be obtained until the middle of November 1944.

A list of the BdO. Verona, dated November 9, 1944, gave the personnel strength of the 15th Police Armored Company as only 60 men. Shortly thereafter the company's strength was listed as five officers, 116 NCO and 29 men, which more or less corresponds to the numbers in the establishment order.

The 15th Police Armored Company may have had the following structure in November 1944:
Leader Group (2 P 40 tanks)
1st Platoon (4 P 40 tanks)
2nd Platoon (4 P 40 tanks)
3rd Platoon (4 P 40 tanks)
Repair group (Workshop Platoon 43)
Replacement crews
Combat and pack train
Motorcycle rifle platoon

Action

The overview of Ordnungspolizei forces and their actions as of July 15, 1944 lists the 15th Police Armored Company in the zone of the BdO. Verona. The same list, dated October 20, 1944, lists the company as being established under the BdO. Verona and serving at San Michele.

On November 11, 1944, 20 armored vehicles were brought, under the leadership of a company officer, from a region south of the Po. The transport command was attacked by low-flying planes, six vehicles were hit several times and the officer was severely wounded.

An overview of the Commander of the Ordnungspolizei under the HSSPF Italy, dated April 9, 1945, listed the 15th Police Armored Company as being stationed in Novarra and having a personnel strength of three officers and 90 men. Its armaments were listed as 13 P 40 tanks, 20 machine pistols, 38 rifles and 80 pistols.

The 15th Police Armored Company was supplied with P40 tanks, the most modern Italian battle tanks, with 7.5 cm tank guns as their main armament. The P40 tank shown above was photographed while with the Army Weapons Office, not the Ordnungspolizei. (BAK)

16th Police Armored Company

Establishment, Structure and Equipment

The 16th Police Armored Company was established according to an order from the Chief of the Ordnungspolizei (O-Kdo. In K (1b) Nr. 34/44(NfD)) of June 28, 1944. The Chief of the Ordnungspolizei referred therein to an order of May 13, 1944 (O-Kdo. In K (Ib) 205 E Nr. 29II/44).

According to the June 28 order, the 16th Police Armored Company, according to War Strength Directive (Army) No. 1149, Version B, was to be established under the Commander of the Ordnungspolizei in Croatia, with 14 armored vehicles taken from the captured vehicles on hand there. This directive, though, served only as a guide, as far as the armored cars were concerned, as the number of forces decreased. The forces for the recovery troop and the division workshop company (mot.) were to be omitted. For the motor vehicle repair group, a workshop platoon with one master and ten NCO and men was to be assigned.

The company was structured, according to the order, as follows:

Leader Group
1st Platoon (2 Skoda and 2 L 6 armored cars)
2nd Platoon (4 L 35 tanks)
3rd Platoon (4 L 35 tanks)
Repair group (Workshop Platoon)
Replacement crew
Combat and Baggage Train

The company's specified strength in personnel, according to the war strength directive, and in view of the smaller crews of the tanks, was three officers, 36 NCO and 59 men.

Only two officers, though, the company chief and leader of the first platoon, were assigned. The Police Motor Vehicle School-Armored Replacement Unit in Vienna was to provide six German NCO (Leader of the company troop, supply train leader, transport leader, equipment manager, bookkeeper and service manager) and one man as a secretary, besides a tank driver, mechanic and armorer for each armored platoon. In case the company needed a German armorer and medic, they could be requested from the Chief of the Ordnungspolizei.

The two tank platoons of the 16th Police Armored Company were equipped with Italian L35 light tanks. Typical of these vehicles were the twin machine guns (here removed) in a ball mantlet on the left side of the body. The crew consisted of driver and gunner. (HH)

Radiomen were not assigned. The other personnel were to be taken from Croatian forces. That meant that the company had only 18 or 20 German police officers and some 70 Croatian volunteers.

Instead of 14 armored vehicles, only 12 were on hand, and of the other needed vehicles, the chief of the Ordnungspolizei could supply only one car and one motorcycle for the leader group, four trucks for the supply train, and one cycle and one truck for each armored platoon.

The assigning of vehicles and drivers, as well as of the workshop platoon, was to be done by special orders. The Police Motor Vehicle School-Armored Replacement Unit was to send the German personnel on the march to the BdO. Croatia on July 17, 1944. The Police Motor Vehicle School (Armored Replacement Unit) also was to supply the armored vehicle crews with clothing. The NCOs and men were each to be supplied with an 08 or 7.65 mm pistol and 80 or 50 bullets, an 84/98 bayonet and a Gas Mask 30 with breathing tube.

The further supplying of the company with weapons as listed in the war strength directive was to be done by the BdO. Croatia. The BdO Croatia was also to report the conclusion of organizing the 16th Police Armored Company to the Chief of the Ordnungspolizei.

The overview of Ordnungspolizei forces and their actions as of July 15, 1944 includes a mine detecting troop with the 16th Police Armored Company, as with the 6th and 11th Companies, which were operating in the same area.

Action

The overview of Ordnungspolizei forces and actions as of July 15, 1944 lists the 16th Police Armored Company with mine detecting troop in the zone of the BdO. Essen. The company was based at Samovar Spa near Zagreb. The overview of Ordnungspolizei forces and actions as of October 20, 1944 shows no changes from July 15. A former member of the company remembers action at the Banya Luka airfield, in which two member of a tank corps and one motorcyclist were killed.

No documentation of the company's further actions is at hand, but it was still listed in the BdO. Croatia's radio plan of April 20, 1945. The company was in Croatia when the war ended. The crews removed the weapons from the armored vehicles and set out in the direction of Germany in their wheeled vehicles.

Training drivers for the small L35 tanks was done by the company. This picture was taken during driver training (note the "Fahrschule" plate on the tank at left). Splinter protection boxes for vehicles have been dug into the small slope in the background. (HH)

Staff Company Armored Platoon, HSSPF., War Zone XVIII

The armored scout-car platoon of the Police Action Staff Southeast was part of the Staff Company of the Police Action Staff, established on January 21, 1942. The armored scout car platoon (see there) left the staff company in the latter half of 1943 and was used to establish the 14th Police Armored Company.

Through the disarming of Italian troops beginning on September 9, 1943, the staff company obtained Italian tanks. On September 19, 1943 it made one (Italian) Panzer 4 and two (Italian) Panzer 10 available for an action of SS Police Regiment 19. The Italian armored vehicles were probably one AB 41 armored scout car (Panzer 4) and two S 37 armored personnel carriers (Panzer 10).

Nothing is known of any further use of the vehicles.

Armored Company with SS Police Regiment 3

On September 20, 1943 the Commander of the Ordnungspolizei in the Netherlands wrote (-VuR. IV. 10.70-) that an armored company was being assigned to the staff of SS Police Regiment 3. and that it was thus intended to direct the unit's finances as of December 22, 1943. Until now, no armored company could be documented in The Netherlands, and all known armored companies were occupied in other places.

Armored Company Center

As 1943 changed to 1944, an Armored Company Center, also called "Light Armored Company Center" could be documented with Police Regiment 13. It is very probable that this company was derived from the armored platoons or the 10th (heavy) Company of Police Regiment Center.

In a war structure report from the "von Gottberg" Battle Group, dated January 10, 1944, the company was listed as having five tanks, three of them with Russian 4.5 cm tank guns (thus Type BA 10) and one Polish tank (7TP).

This company existed in addition to the police armored companies established centrally in Vienna at that time, and was not included in their numbering.

Horsedrawn Ordnungspolizei units advance on a Russian village in 1943, along with a 7 TP tank of Armored Company Center. The tank is marked with German crosses on its turret. (BAK)

Armored Car Platoon with SS Police Regiment 1

In an announcement of the Inspector-General of the Police Troops on September 28, 1943 (Gr. Org. Nr. 1459743 g.Kdos.) about the future use of captured Italian tanks by the Army Group Southeast, the equipping of four SS Police Regiments ·with one armored reconnaissance platoon each, with five Italian tanks, was mentioned. SS Police Regiments 1, 2, 14 and 18 were listed for the first rate of establishment. In a further plan from the Inspector-General of the Armored Troops, also of November 1943, about units of Army Group Southeast that were equipped with captured Italian armored vehicles, their equipping is ordered or, if they were soon to be established, likewise four SS Police Regiments and their equipping with a total of 20 L6 tanks were noted. It is not now known when and if these establishments and equippings took place. In a troop structure report of the General Command, XXXIV. Army Corps, of December 15, 1944, a Police Armored Platoon 1 is listed along with the Engineer Platoon and Intelligence Platoon of Police Regiment 1 in the Police Security Zone South. Unfortunately, no further data on the makeup and equipment of this unit has been found.

Armored Car Platoon with SS Police Regiment 2

In the records of the subsequent use of captured Italian tanks by the Army Group Southeast after the end of September 1943 (see Armored Car Platoon, SS Police Regiment 1), the planned establishment of an armored car platoon for SS Police Regiment 2 is also mentioned.

No documentation of an armored car platoon with SS Police Regiment 2 has been found.

Armored Car Platoon with SS Police Regiment 14

In the records of the subsequent use of captured Italian tanks by the Army Group Southeast after the end of September 1943 (see Armored Car Platoon, SS Police Regiment 1), the planned establishment of an armored car platoon for SS Police Regiment 14 is also mentioned.

No further documentation of an armored car platoon with SS Police Armored Regiment 14 has been found.

Armored Car Platoon with SS Police Regiment 18

In the records of the subsequent use of captured Italian tanks by the Army Group Southeast after end of September 1943 (s. Armored Car Platoon, SS Police Regiment 1), the planned establishment of an armored car platoon for SS Police Regiment 18 is also mentioned.

According to information from a former member of Armored Car Platoon 18, the unit was equipped with Italian tanks in Athens in September 1943. About a dozen tanks were on hand in Athens and were used by driving teachers for driving school lessons. Weapons training was done by the troop itself. This ex-member remembers seeing service with the Armored Car Platoon of Police Regiment 18, equipped with four tanks. In the retreat from Greece, the tanks were shipped out of Saloniki and transported to Bulgaria. In the end, only two tanks, both with war damage, remained. In February-March 1945 the unit again had two tanks, also Italian types; they remained with the troops till they reached Maisburg on the Drau and were surrendered there.

The overview of Ordnungspolizei forces and their actions as of July 15, 1944 lists an armored car platoon with SS Police Regiment 18, quartered in Athens. The command chart of the Commander of Greece lists an armored reconnaissance platoon with armored scout cars (2 cm guns) and five assault guns (4.7 cm guns) with SS Police Regiment 18. The overview of Ordnungspolizei forces and actions as of October 20, 1944 likewise lists an armored car platoon with SS Police Regiment 18, then in the zone of the BdO. Serbia.

At the end of November 1944, the 118th Jäger Division, in service on the Syrmia Front north of Vukovar, reported on SS Police Regiment 18, then subordinate to it. The total manpower of the regiment was only about 10%. Of the specified strength of the regiment, with five 4.7 cm assault guns and three scout or reconnaissance cars with 2 cm guns, only one assault gun was present. The fuel for it, if no new supplies were delivered, was only enough for ten kilometers, after which the vehicle had to be destroyed.

On May 23, 1944 a parade of Wehrmacht and police troops serving in Greece was held before the Military Commander of Greece, Air Force General Speidel/ Here two L6 tanks of SS Police Regiment 18 are passing the military commander. (BAK)

An assault gun crew of the armored car platoon of SS Police Regiment 18 busies itself in the summer of 1944 by reading the newspaper "Die Deutsche Polizei". The men of Assault Gun 184 wear tropical uniforms with shorts and pith helmets.

Reinforced Police Armored Car Platoon Berlin

The overview of Ordnungspolizei forces and their actions as of October 20, 1944 lists a strengthened Police Armored Car Platoon under the Police Commander of Berlin, with quarters in Berlin.

No further documents tell of the establishment or equipment of this platoon. Photos from the Reich Chancellery, taken in May 1945, show an old Daimler armored car with police registration, and behind it a Krupp (Netherlands) armored car. It is presumed that these two vehicles belonged to the reinforced Police Armored Car Platoon.

Two vehicles of the reinforced Armored Car Platoon of the Berlin Police, seen in the Chancellery yard in May 1945. In front is an old Daimler DZVR with police registration and German crosses painted on; behind it a Krupp armored car from The Netherlands.

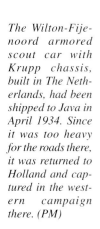

The Wilton-Fijenoord armored scout car with Krupp chassis, built in The Netherlands, had been shipped to Java in April 1934. Since it was too heavy for the roads there, it was returned to Holland and captured in the western campaign there. (PM)

Armored Car Platoon with SS Police Regiment Bozen

On October 1, 1943 the Police Regiment South Tyrol was established in Bozen; on October 29 it was renamed Police Regiment Bozen. From April 1944 on the unit was designated SS Police Regiment Bozen and added a fourth battalion. The I./SS Pol.Rgt. Bozen saw action in the Abazzia and Fiume areas of Istria as of May 1944. Photos show that the First Battalion had an armored car platoon with two Italian vehicles, an old Lancia built in 1919 and a Fiat Ansaldo AB 41. The photos show that the battalion saw action in the Pola area (Pola-Marzana Road), Fiume, on the Trieste-Abbazia and Santa Lucia-Isonzo roads from June 1944 to the beginning of 1945. The AB 41 armored car was also loaded onto a transport ship, probably for use against the islands along the coast.

The same photos show at least two armored cars with I./SS Pol.Rgt. Bozen, or being used by that unit.

Armored Company with SS Police Regiment 15

In the directives from the police administrative officials to the police units serving in Italy on taking positions, an armored company with an administrative officer is listed for the I./SS Pol.Rgt. 15 from February to November 1944. It could not be determined just which company it was.

The First Battalion of SS Police Regiment 15 had at least one Lince (I) armored scout car in April 1945. On April 21, 1945 this scout car saw action on the Romagnano-Novarra road. Here a supply vehicle carrying 15 men hit a mine and was attacked by some 180 bandits. The action of the armored scout car and its two-man crew was sufficient to drive the bandits away, rescue the surviving soldiers, remove all the wounded and dead and tow the truck away.

The two armored cars used by the Police Volunteer Regiment of Bozen; in the rear is an old 1919 Lancia, armed with two machine guns, and in front a relatively modern AB 41, with a 2 cm gun and two machine guns. (RE)

The AB 41 armored scout car is leading a column with Mlkw.Pzkw. They were often used as security vehicles to accompany transport columns. (RE)

The AB 41 armored car is camouflage-painted in typical Italian war zone style, in yellow with brown and green stripes. The opened turret hatch and upper part of the left side door can be seen clearly. (RE)

The Lancia armored car is seen on a mountain road with a picturesque view of a valley. This vehicle is also typically camouflaged in the style also used by the Wehrmacht. (RE)

At least two small L 33 armored cars were used by the Police Regiment Bozen. This newly delivered Type L 33/II vehicle is still marked "I./Pol. Bozen" (1st Battalion, Police Regiment Bozen). The officer wears a tropical uniform with shorts. (BAK)

Armored Company with SSPF. Adria West

In the war structure report of the HSSPF. Italy of April 1945, a half armored company is listed for the SSPF Adria West. This was subordinated or assigned to the Grenadier-Jäger Brigade of the SS (Karstjäger). In a report of the 1st Derbyshire Yeomanry of New Zealand on combat in the Gemona-Ospadalleto area in late April and early May 1945, the shooting down of two P40 tanks was noted and documented by photos. Since the P40 tanks were used only by the police, this could indicate an unknown police armored company, since the Police Armored Companies 10 and 15, equipped with P40 tanks, were not in that area at that time.

At the beginning of 1945, the Waffen-Grenadier-Jäger Brigade of the SS (Karstjäger) had an armored company subordinated or assigned to it. It is not known whether this company was set up as a police armored car company.

Soldiers of the New Zealand 1st Derbyshire Yeomanry sit on the rear of a P40 tank shot down in the Gemona-Ospadalletto area at the end of April 1945. The company to which this tank belonged was assigned or subordinated to the SS Mountain Jäger Division (Karst-jäger). (IWM)

The horseshoe on the track apron at the left front does not seem to have brought the P40 or its crew much luck. Officers of the 1st Derbyshire Yeomanry study a map beside the vehicle, which may have belonged to a police armored company. (IWM)

CHAPTER IV
TRAINING AND REPLACEMENT UNITS

General Information

In the following sections, an overview of the training and studies of the Ordnungspolizei personnel specially trained for service with armored vehicles will be offered. Along with this personnel, who were specially trained for armored vehicles, other personnel naturally came to serve in the armored units too, such as auto drivers, motorcyclists, motorcycle riflemen, intelligence personnel, repair-shop personnel, medical personnel, etc. These men were also trained in various schools and replacement units and then assigned to the new armored units or the Ordnungspolizei's armored school (later the Police Armored Replacement Unit).

Some examples are the Motor Vehicle Replacement Unit in the Police School for Technology and Traffic in Berlin (car, truck and motorcycle drivers), the Police School for Technology and Traffic in Berlin (workshop personnel), the Police Motor Vehicle School in Dresden (drivers, workshop leaders), the Driver Replacement Unit of the Police Motor Vehicle School in Iglau (car, truck and motorcycle drivers), the Police Intelligence Replacement Company in Krakau (intelligence personnel), and the Police Medical Replacement Unit in Berlin (medical personnel). Police marksmen were trained and made available by the Police Training Battalions (later renamed Police Weapons Schools) I, II, III and IV.

The personnel trained on armored vehicles returned to their bases in Germany, from which they had been sent for training, when their training was finished, unless they were to be used to set up an armored unit right after their training. At their bases they served as usual, unless they were called on to join an armored unit in times of need. Only after the establishment of the Police Armored Replacement Company at the end of April 1943 were trained personnel, except for officers, assigned to that unit.

Provision of replacement officers was carried out by a "Leader Reserve" created at the Ordnungspolizei Headquarters in mid-1943.

The police troop units in action were to request replacements only in cases of death. This policy was departed from only when, for example, a shortage of at least two men was expected because of wounds or illness. Replacement of NCO and men was only to be requested when there had been a loss of more than 10% of the unit's specified strength.

Wounded or sick personnel, after being released from the hospital, returned to their German bases, where they were granted appropriate recovery time or a year's leave.

After the leave, the personnel were returned to their replacement unit. From there they were generally sent back to their police troop unit.

The Police School for Technology and Traffic, Berlin

In the 1920s and 1930s, it was only the drivers among the crews of police vehicles and other special vehicles who were given centralized training. They underwent a four-week course at the Police Institute for Technology and Traffic in Berlin (later renamed the Police School for Technology and Traffic, Berlin). Two drivers and two reserve drivers had to be trained for every special vehicle. The participants were sent by the individual states or police headquarters and came exclusively from the motor vehicle services. As a rule, several courses were offered every year.

The training of the rest of the crews took place decentrally in the locations of the special vehicles and differed from state to state and base to base. Since the heavy machine gun was the heaviest weapon used in the special vehicles, it was chiefly machine gunners who were assigned to the crews at the bases. Commanders of the special police vehicles were longtime NCOs or officers in the motor vehicle services. The cooperation of the crew was tested in special training trips and classes.

In April 1936 (RMBliV. p. 505) a six-week training course for drivers of special vehicles was developed at the Police School for Technology and Traffic. Nominations for this course were to reach the RuPrPdI. by September 1, 1936. Whether this course was given is not known.

No courses at the Police School for Technology and Traffic after 1936 are known from the available sources. At first, trained crews from the former Austrian and Czech Gendarmerie and the Viennese Security Watch were available for the Steyr, Skoda and Tatra armored cars taken over from 1938 on. Training for armored car drivers and crews was given from 1939 to 1941 by the Driving Command of the Vienna Police and the units involved.

Documents and pictures have been found, though, that show work done on Steyr armored cars at the Police School for Technology and Traffic in Berlin. Whether this was done to train drivers and crews or just for technical developments is not known.

The Police School for Technology and Traffic in Berlin was responsible for the training of officers, NCOs and men involved in motor vehicle, transport and intelligence communications. In addition, the training of armorers and officials for the motor vehicle units took place there. The Police School for Technology and Traffic was also the central supply point for motor vehicles and equipment, as well as weapons and supplies. It was simultaneously the central maintenance depot for large weapon repairs.

Armored Replacement Unit 100 in Schwetzingen

Since the Ordnungspolizei scarcely had any experience with tanks (except a few Polish 7 TP), and had no captured French tanks in their possession, before the end of 1941, training personnel and crews for these vehicles first had to be trained by the Wehrmacht. This was done on orders from the Reichsführer-Chief on December 17, 1941, by Armored Replacement Unit 100 of the Wehrmacht in Schwetzingen. This replacement unit had been established in April 1941 for the training and replacement preparation of personnel for the armored units of the Wehrmacht equipped with captured armored vehicles.

According to the Reichsführer-Chief's order of December 17, 1941 (O-Kdo. I K (2) 252 Nr. 118/41), with the approval of the OKH, Chief H Rüst u. BdE.-AHA Ag K/In 6 (III E) Nr. 4280/41, of November 28, 1941, an armored vehicle training course for officers, masters and policemen (SB) of the Ordnungspolizei was given at Armored Replacement Unit 100 in Schwetzingen, beginning on January 7, 1942. The participants in this training were to report in Schwetzingen on January 6, 1942.

Special vehicle drivers were trained at the Police Institute for Technology and Traffic, later renamed the Police School for Technology and Traffic, in Berlin. A Benz 21 special vehicle is seen driving through a gate at the school during driver training. (BAK)

Taken during a visit of Danish police officers to the Police School for Technology and Traffic in Berlin. It has not been determined whether Steyr armored cars were used by the driving school as well as for testing purposes. (BAK)

Training with a Hotchkiss H38 driving school tank went on at Armored Replacement Unit 100 in Schwetzingen in January 1942. It is interesting that some of the policemen wear Wehrmacht uniforms. The reason for this is not known. (RC)

Assigned were:

a) Officers and NCO for training in tank driving courses
b) NCOs for training in armored car maintenance
c) Officers and NCO for training as tank drivers.

The training for a) and b) lasted through February 17, 1942, that for c) only until January 27, 1942. When the training ended, the officers and men returned to their regular bases.

A special order regarding their further utilization was to follow.

Food and housing for the duration of the training were provided at Armored Replacement Unit 100 in return for reimbursement.

All further training on armored vehicles, designated as tanks by the Ordnungspolizei, took place at the Police Motor Vehicle School in Vienna.

Police Motor Vehicle School, Vienna
Only with the intended establishment of larger armored forces (units and companies) and the increase in numbers and varied types of armored vehicles was a central training school for drivers of armored vehicles created. This was at the Ordnungspolizei's Motor Vehicle School in Vienna, and was set up by an order of October 22, 1940 (MBliV. S. 1995).

The earliest extant document that refers to a planned establishment of armored units and central training for them by the Ordnungspolizei is an order from the Reichsführer SS and chief of the German Police of September 15, 1941 (O-Kdo. 1 K (2) 252 Nr. 79/41):

It states that in the very near future a number of new armored vehicles will be delivered to the Ordnungspolizei, and that it is thus necessary to begin training drivers for these vehicles at once. The training

Off-road driving training on a Steyr armored car took place at the Police Motor Vehicle School in Vienna in November and December 1941 (1st armored vehicle course). The trainees are wearing normal coats, some with leather hoods for drivers. (WP)

should take place at the Motor Vehicle School in Vienna. Only such police officers who had a level 2 driving license for a long time and shown outstanding technical driving skills were to be chosen for training. If possible, they should not be taller than 1.75 meters, on account of the seating conditions in the armored vehicles.

During their armored vehicle training, the men were to learn not only driving, but also things like disposing of obstacles. Since the extent of the course could not disregard the small numbers of training vehicles, the men were simply to be registered with the police headquarters for the time being, so that they could be summoned to the school at any time. Two months later, the order of September 15 was overridden or changed by the order of the Reichsführer SS-Chief of the German Police on November 7, 1941 (O-Kdo. I K (2) 251 Nr. 186/41), concerning the establishment of armored units and the training of crews. Accordingly, the drivers to be placed in crews were to be ordered to the Motor Vehicle School in Vienna for a training course beginning on November 19, 1941, by order of the Reichsführer-Chief of November 12, 1941 (O-Kdo.I K (2) 251 Nr. 186II/14).

In addition, men who were to be trained in armored vehicle maintenance were to be sent to this training course. The men were to arrive at the Motor Vehicle School in Vienna by 7:00 P.M. on November 18, 1941. Their training was intended to continue until December 15, 1941. From that point on, the men, with their order to the Motor Vehicle School being lifted, were to report to the police headquarters in Vienna, where weapons training for the whole crews was to be carried out until the end of January 1942.

The following training course for drivers of armored vehicles took place at the Motor Vehicle School in Vienna:

11/19/41-12/15/41: Training in armored vehicle driving and maintenance. "1st Armored Car Course".

By order of the Reichsführer-Chief on March 6, 1942, the Police Motor Vehicle School in Vienna was renamed the Driving and Armored Vehicle School of the Ordnungspolizei in Vienna (see there).

Motor Vehicle Echelon Center of the Police Headquarters, Vienna

The weapon training of the armored vehicle crews, including drivers and courier motorcyclists (the armored cars originally had no radio equipment) was carried out in special courses at the Motor Vehicle Echelon Center (Armored Vehicle Readiness) of the Vienna Police Headquarters.

This was in the Otto Steinhäusl Barracks (ex-Moroccan Barracks) on the Marokkanergasse in the Third Ward of Vienna. The establishment and appointment of suitable crews was done by the most varied police authorities, which were charged to do so by order of the Reichsführer-Chief in the Ministry of the Interior. The crews were trained to operate the vehicles, their mounted weapons and hand weapons. For example, training on the tank gun (2 cm Kwk 35), MG 34, MP 34 and Stick Grenade 24 was typical.

The following courses for armored vehicle crews were conducted there:

12/15/41-1/??/42	"1st course for crews of Steyr armored cars"
3/3/42-3/31/42	"2nd course for crews of Steyr armored cars"
4/10/42-5/15/42	"3rd course for crews of Steyr armored cars"
6/15/42-7/24/42	"4th course for crews of Steyr armored cars"

The fourth course for crews of Steyr armored cars was originally to be given from 5/29/42 to 7/10/42. This course, though, was used for a special term from May 29 to June 12 and unified into an "Armored Car Unit" (see there). The 4th armored car course (training for crews of Steyr armored cars) thus took place only from June 15 to July 24, 1942.

Twelve brand new second-series Steyr armored cars are parked in the yard of the Otto Steinhäusl Barracks in Vienna. The first training courses for crews of Steyr armored cars were given in them, and later the first armored companies were supplied with them. The crosses painted on the turrets and front, side and rear bodies are easy to see. (AW)

Vehicles of the first training course for Steyr armored car crews are seen in the yard of the Otto Steinhäusl Barracks on Marokkanergasse in Vienna. The course was conducted by the Motor Vehicle Echelon Center of the Vienna police in the winter of 1941-42. (WP)

When possible, only those Wachtmeister (SB) were to be sent from German bases as armored car commanders and gunners who had already completed machine-gun training. If necessary, machine-gun training was to begin at once. The men intended as armored car commanders could not be more than forty years old and, when possible, were to be masters. Otherwise, capable Hauptwachtmeister, no more than forty years old and eligible for promotion to Master, could be assigned. Only Wachtmeister (SB) who had already had an armored car training course at the Police Motor Vehicle School in Vienna could be assigned by their home bases to the course to be trained as armored car drivers and mechanics.

The participants had to arrive at the Motor Vehicle Echelon in the Otto Steinhäusl Barracks, Marokkanergasse, Third Ward, Vienna, by 7:00 P.M. on the day before the course began.

All participants were officially housed and fed by the Vienna Police Administration. They were to be supplied only with pistols. The participants in the course were to be supplied with service clothing by their home bases according to Section 27, 3 Pol.Bekl.V. Part II.

The economic conditions stated in the order of October 28, 1941 (RMBliV S. 1956 E) applied to these courses as well.

Motor Vehicle and Armor School of the Ordnungspolizei in Vienna

By order of the Reichsführer-Chief in the Ministry of the Interior on March 6, 1942 (O-Kdo. I O (3) 2 Nr. 333/42), the "establishment of an armored car school of the Ordnungspolizei and renaming of the Police Motor Vehicle School of Vienna" were called for. The increased need for qualified central training of armored vehicle personnel and the central establishment of armored units were recognized. The order stated:

1. Effective 3/1/1942, the Police Motor Vehicle School of Vienna is to include an armored car school with armored car equipment office.
2. The armored car school and equipment office are to be located until further notice at the quarters of the large Gend. Company (mot.) in Vienna-Purkersdorf.
3. The economic conduct and supplying will be done by the Police Motor Vehicle School of Vienna.

4. The determination of the temporary structure and assignment of the necessary personnel and vehicles will be done by special order.

5. Effective immediately, the Police Motor Vehicle School of Vienna will be designated "Motor Vehicle and Armored Vehicle School of the Ordnungspolizei in Vienna".

6. The <u>Order of 10/22/1940 (MBliV. S. 1995)</u> is to be applied accordingly.

7. Because of the address etc. of the Motor Vehicle and Armored Vehicle School of the Ordnungspolizei in Vienna, the aforementioned order applied.

Thus an independent training unit for the Ordnungspolizei's armored troops was created, and with the attached armored vehicle equipment office, it was also able to supply these troops. The first courses given by the "Motor Vehicle and Armored Vehicle School of the Ordnungspolizei" were:

5/19/42-6/27/42   "Training course for crews of armored cars (Hotchkiss)"

7/7/42-8/1/42   " Driving school training for wheeled armored car vehicles over ten tons (Steyr armored scout cars)"

For the training course for crews of Hotchkiss tanks, tank mechanic nominees were to be Wachtmeister (SB) who had already had a workshop leader class at the Police Motor Vehicle School in Dresden, or if this was not possible, a driving class at the Motor Vehicle and Armored Vehicle School in Vienna. Within the crews, one of the available inspectors (commanders) of the Ordnungspolizei was to be assigned as the platoon leader. The men intended to be platoon leaders could not be over forty years old and should, if possible, be masters. Otherwise, suitable Hauptwachtmeister, no more than forty years old, who were about to be promoted to Master could be assigned as platoon leaders.

The men to be sent to the training course were to be examined in advance at their home bases, to make sure their health could stand the rigors that armored car crews faced. Reference was made to the instruction that these men, if possible, should not be over 1.75 meters tall and should have a Class 2 driver's license. All participants (including tank mechanics) were to be trained as tank drivers and leaders. The course was to be run according to HDV. 470/5a,/5b, 470/6 and 470/9. Special orders concerned school shooting.

In an order of July 15, 1942 (O-Kdo.I K (2) 252 Nr. 110/42), the Reichsführer-Chief mentioned that the establishment of the motor vehicle and armored vehicle school of the Ordnungspolizei in Vienna was finished, and all training courses for the crews of armored vehicles could be offered there. Parallel to the establishment of the school, the Vienna Police, Driver Echelon Center, also offered training courses.

In a directive of July 21, 1942 (O-Kdo. I-Ia (1) 2 Nr. 139/42), new regulations were given for providing replacements for police troop units in service outside Germany. As of August 15, 1942 all armored car and tank companies and platoons received replacements from the Motor Vehicle and Armored Vehicle School of the Ordnungspolizei in Vienna.

The following further courses can be documented to date:

8/12/42-9/25/42   "Training course for crews of Steyr armored cars", plus "Training as armored platoon leaders for 5 lieutenants of the school" and "Completion of training as leaders of armored companies for 3 captains of the school"

10/1/42-10/17/42   "Training course for drivers of Panhard armored cars"

10/19/42-11/7/42   "Training course for crews of Panhard armored cars"

1/6/43-1/31/43   "Training course for drivers of Panhard armored cars"

2/1/43-2/13/43   "Training course for crews of Panhard armored cars"

2/15/43-3/6/43   "Training course for drivers of ???"

The participants always were to report to the Motor and Armored Vehicle School, Landstrasser Hauptstrasse 68, Vienna by 7:00 P.M. on the day before the course began. The course from 8/12/42 to 9/25/42, for example, had 70 participants. Wachtmeister (SB) who had already had a tank-driving course at the Motor and Armored Vehicle School of the Ordnungspolizei in Vienna could train to be tank drivers and mechanics. The men to be trained as tank leaders could not be over 42 years old and, if possible, were to be Masters, or at least capable Hauptwachtmeister no more than 42 years old who were due to be promoted to Master. The men to be sent for training were to be checked at their home bases to see if they were healthy enough to stands the rigors of tank-crew work.

All the participants were to be housed at no cost and fed on payment at the Motor and Armored Vehicle School of the Ordnungspolizei. The participants in the course were to be supplied with clothing for outside service, plus anything else they needed, according to the order of July 12, 1941 (MBliV. S.1303). They were likewise to be given a Pistol 08 or 7.65 mm and fifty rounds of ammunition. Special clothing, as listed in Section M, Appendix 1, Police Clothing Prescription, Part II, was to be provided for the duration of the course by the Vienna Police Administration.

The economic conditions called for in the order of 10/28/41 (RMBliV S. 1956 e) were also to apply to these courses. It was stressed that the order of 7/18/1939 (MBliV. S. 1535) noted in the aforementioned order had been replaced by the order of 5/21/1942 (MBliV. S. 1078).

For shooting training on the Steyr armored car, the Technical Police School-Arsenal of Berlin drew on the supplies of the Subarsenal of Vienna to supply each tank gunner with 6 tank shells with tracers for Kwk. 35 and 75 machine-gun bullets.

For the driving course, Wachtmeister of the School (SB) with Class II drivers' licenses for special technical driving were to be assigned.

The growing training tasks and the growth of armored police units led at the end of 1942 to the renaming of the school as the Police Armored Replacement Unit of Vienna.

The Police Armored Replacement Unit of Vienna

By order of the Reichsführer-Chief in the Ministry of the Interior on December 19, 1942 (O-Kdo. I O (3) 1 Nr. 398/42), the "establishment and structuring of a police armored replacement unit at the Police Motor Vehicle School in Vienna" was ordained. The order read:

1. The armored vehicle school with armored equipment office, established at the Police Motor Vehicle School in Vienna according to the order of 3/6/1942 (MBliV. S. 512), will be renamed "Police Armored Replacement Unit of Vienna" (Pol.-Panz.-Ers.-Abt.).

2. The Police Armored Replacement Unit belongs to the Police Motor Vehicle School of Vienna and is directly subordinate to the commander of this school.

3. The unit commander of the Police Armored Replacement Unit is hereby granted the powers of a non-independent battalion commander, according to Section 9 of the current Service Orders for Police Troops.

4. (1) The Police Armored Replacement Unit is a training and replacement unit for police armored units. (2)Personnel replacements needed by the police armored units shall be requested from the police regiments or directly from the police Motor Vehicle School in Vienna.

5. The structure of the Police Armored Replacement Unit is the express concern of the Police Motor Vehicle School in Vienna.

6. (1) The order of 3/6/1942 (MBliV. S. 512) is hereby canceled. (2) Because of the significance of the Police Motor Vehicle School in Vienna, I refer to the order of 8/5/1942 (MBliV. S. 1640).

The structure referred to under Section 5 of the order of 12/19/1942 applied to the Police Motor Vehicle School with the order (O-Kdo. I O (3) I Nr. 398/42) of 1/7/1943. This order, unfortunately, could not be found to date, so that the structure is not known. But there is an order of the Reichsführer-Chief (O-Kdo. II P II (2b) 56 c Nr. 65/43) of 3/1/1943 that regulates the assignment of personnel to occupy positions in the Police Armored Replacement Unit. According to it, cadre personnel for the Police Armored Replacement Unit to be established at the Police Motor Vehicle School in Vienna, effective immediately, were:

One Hauptmann of the Schutzpolizei, two Oberleutnante, one Leutnant, two Revier-Leutnante, six Meister, two Zugwachtmeister, four Revier-Oberwachtmeister and one Oberwachtmeister, all of the Schutzpolizei.

From the Vienna Police Headquarters, three Unterführer (only Wachtmeister to Zugwachtmeister inclusive) were to be assigned to the Police Armored Replacement Unit. Police reservists could also be assigned for this purpose.

The Unterführer of the cadre personnel were regarded as transferred in reference to service clothing. The officers, Revier officers and Unterführer were each to be supplied with a 7.65 mm pistol with 50 rounds, an 84/96 sidearm (except officers and Revier officers), and a gas mask by their home bases.

Driving school training with a Renault tank on the road between Purkersdorf and St. Pölten. To save fuel by cutting the weight, the turrets were removed for driver training. This tank was later used by the 7th Police Armored Company. (RS)

Hotchkiss and Renault tanks in the yard of the Motor Vehicle School of the Ordnungspolizei in Vienna-Purkersdorf. The school was located in a former Gendarmerie barracks, but because of space limitations it also used private quarters such as small hotels and parking areas. (RS)

Another Renault tank in the yard of the Motor Vehicle School in Purkersdorf. At its left is the Renault with registration number 20888, fitted with a cover to keep dust out. (JH)

Thus the Police Armored Replacement Unit was supplied with personnel and could commence its training service. The following courses are known to have been given:

4/13/43-5/3/43	"Course for crews of (Steyr) armored scout cars", driving technology section.
5/4/43-5/22/43	"Course for crews of (Steyr) armored scout cars", weapons training for the crew.
5/25/43-6/3/43	"Course for crews of Hotchkiss and Renault tanks", driving technology section.

For the armored scout car training courses of 4/13-5/22/43, two officers and 13 NCO and men were assigned to the driver training section, while 25 NCO and men took part in the weapons training for tank gunners. This amounts to the training of crews for six Steyr armored scout cars.

In addition, three tank radiomen, already on hand, were to be trained in this course by the Police Motor Vehicle School (Armored Replacement Unit) in Vienna.

The officers, NCO and men to be assigned to this training were likewise to be checked by their home bases to see whether they were equal to the burdens imposed on armored car crews. All the NCO and men could not stand over 1.75 meters tall, and could not have been born before 1901. Their training was to take place according to the existing guidelines. The participants in the course were to be supplied by their home bases with clothing for outside service, according to the order of 7/12/1941 (MBliV. S.1303) and the later additions to it. Special clothing, as in Section M, Appendix 1, Police Clothing Instructions, Part II, was to be supplied for the duration of the course by the Police Motor Vehicle School in Vienna. The economic requirements stated in the order of 7/10/1942 also applied to these courses.

The home bases were to supply a Pistol 08 or 7.65 mm with 64 or 50 rounds of ammunition per man.

For shooting training, the Police School for Technology and Training-Arsenal-in Berlin was to supply 75 machine-gun cartridges for each tank gunner.

Shortly after this training, on April 29, 1943, Order O-Kdo. I K (1b) 251 Nr. 168/43 commanded the establishment of a Police Armored Replacement Company of the Police Armored Replacement Unit, as this has already been planned for in the (unknown!) structure plan of January 7, 1943. To form the planned Police Armored Replacement Company of the Armored Replacement Unit, an unknown number of NCOs and men was thereby ordered to the Police Motor Vehicle School. On July 8, 1943 a further order (O-Kdo. I K (1b) 251 Nr. 286/43) was given to the Police Armored Replacement Company. By it, 44 NCOs and men were assigned to the Police Armored Replacement Company to fill the Armored Replacement Unit, as per O-Kdo. I K (1b) 251 Nr. 168/43 of April 29, 1943. In both orders it was stated that the home bases were to provide the following pieces of clothing and weapons:

A field cap, a field blouse of earlier or current type, long trousers, a coat, a pair of heavy laced shoes, a pair of leggings, a neckerchief, three shirts, three shorts, three pairs of socks and a body belt with pouch and lock. Also a Pistol 08 or 7.65 mm with 80 or 50 rounds and a Sidearm 84/98.

The 44 NCO and men were to report to the Police Motor Vehicle School, Landstrasser Hauptstrasse 68, in Vienna by 7:00 P.M. on July 20, 1943.

In January 1944 the Police Armored Replacement Unit once again was informed of a change in its structure. This was done by order of the Reichsführer-Chief (O-Kdo. I Org. (3) 1 Nr. 511/43) of January 5, 1944. Effective immediately, the following structure applied to the Police Armored Replacement Unit with its cadre personnel:

Staff of the Police Armored Replacement Unit
One Major of the Schutzpolizei as Unit Commander, two officers, one Revierleutnant, plus twelve other NCO and men.

Training Company
One Hauptmann of the Schutzpolizei as Company Leader, two Oberleutnante of the Schutzpolizei as Platoon Leaders and Instructional Officers, one Revierleutnant of the Schutzpolizei as Workshop Leader, plus 23 other NCO and men.

Replacement Company
One Hauptmann of the Schutzpolizei as Company Leader, three Oberleutnante of the Schutzpolizei as Platoon Leaders, plus six other NCO and men.

Intelligence Company
One Hauptmann of the Schutzpolizei as Company Leader, one Oberleutnant of the Schutzpolizei as Platoon Leader and Instructional Officer, plus 11 other NCO and men.

Instructional Company
One Hauptmann of the Schutzpolizei as Company Leader, four Oberleutnante of the Schutzpolizei as Platoon Leaders, plus 12 other NCO and men.

Armored Equipment Office
One Revierleutnant of the Schutzpolizei as Equipment Office Leader, plus 21 other NCO and men.

Armored Vehicle Workshop
One Revierleutnant of the Schutzpolizei as Workshop Leader, plus 12 other NCO and men.

Motor Vehicle Echelon
One Oberleutnant of the Schutzpolizei as Echelon Leader, one Revierleutnant of the Schutzpolizei as Motor Vehicle Service Leader, plus 16 other NCO and men.

The Police Armored Replacement Unit thus had a complete strength of cadre personnel of 137 persons (18 officers, 5 Revier officers and 119 NCO and men). The assignation of still lacking motor vehicle technical personnel and motor vehicles was to be done by special order, as was the assigning of intelligence personnel.
The following motor vehicles were on hand:

a) Motor Vehicles: b) Armored Vehicles:
3 Pkw 4 staff cars   6 Panhard armored cars
3 Stkw. 4 personnel carriers   1 Steyr armored car
2 Sstkw. 8 personnel carriers   4 Renault tanks
(Command radio cars)4 Hotchkiss tanks
2 Wkw. 1 supply trucks   2 BT 7 tanks
6 Wkw. 3 supply trucks   1 BT 5 tank
10 Motorcycles with sidecars   1 T 26 tank
15 Motorcycles   1 T 34 tank

After this restructuring, the following additional courses were announced:

3/15/44-4/11/44   "Course for crews of armored scout cars (Steyr)", driving technical part,
4/12/44-5/20/44   "Course for crews of armored scout cars (Steyr)", weapon technical part,
6/27/44-7/24/44   "Course for crews of tanks (Renault) and armored cars (Steyr)", driving technical part,
7/25/44-9/23/44   "Course for crews of tanks (Renault) and armored cars (Steyr)", weapon technical part.

For these courses, two Oberleutnante and two Masters were transferred from their home bases for driving technical training as platoon and tank leaders as of 3/15/1944, ten NCOs and men with drivers' licenses from the Driver Replacement Unit of the Police Motor Vehicle School in Iglau as tank drivers, plus 12 NCOs and men with Class 1 drivers' licenses as motorcycle drivers. Also, five tank radiomen were assigned from the Armored Replacement Unit of the Police Motor Vehicle School in Vienna, and three NOCs and nine men from Police Weapons School III in The Hague as motorcycle (machine gun) gunners.

For the weapons training that followed the driving training for the whole crews from 4/12 to 5/20/1944, five more NCOs as tank leaders and 20 NCOs and men as tank gunners were assigned from Police Weapons School III in The Hague. For the course that began on 6/27/1944, 12 Leutnante of the Schutzpolizei from various bases and one Hauptwebel of the Schutzpolizei from the Police Motor Vehicle School in Iglau, Driver Replacement Unit were assigned for driver training, 44 NCOs and men with Class 2 drivers' licenses as motorcycle drivers, plus five NCOs and 15 men from Police Weapons School I in Hellerau as motorcycle (machine gun) gunners.

For the weapons training for the entire crews following the driver training from 7/25 to 9/23/1944, 30 more NCOs and men were assigned from Police Weapons School I in Hellerau as tank gunners.

The officers, NCOs and men assigned to these courses were to be checked by their home bases to make sure they could stand the pressures on armored vehicle crews. All NCOs and men could not be more than 1.75 meter tall and not be born before 1905.

The participants in the courses were to be supplied by their home bases with the following new-type clothing and weapons.

One field cap, one field blouse of former or new type, long trousers, one coat, a pair of heavy laced shoes, a pair of leggings, one neckerchief, three shirts, three shorts, three pairs of socks and one body belt with pouch and buckle. Also one 08 or 7.65 mm pistol with 80 or 50 rounds and one 84/98 sidearm.

Special clothing according to Section M, Addendum 1, Police Clothing Directive, Part II, was to be provided by the Police Motor Vehicle School in Vienna for the duration of the course.

The economic specifications stated in the order of 7/10/1942 (RMBliV. S. 1446) also applied to this training. The courses were to be held according to Army regulations.

For gunnery training, the Police School for Technology and Traffic-Arsenal in Berlin was to provide, from the supplies at the Sub-Arsenal in Vienna, six 2 cm antitank shells and 75 rounds of machine-gun bullets for each tank gunner for the Steyr vehicle, and 75 rounds of 7.5 mm S (f) for machine guns (f) and six 3.7 cm 147 (f) tank shells for each gunner for the Renault tank.

The NCOs and men assigned to the courses were assigned, with the exception of the officers, to the police Armored Replacement Unit when the training was finished.

Further documentation of assignment to courses at the Police Armored Replacement Unit for armored car and tank crews and motorcycle riflemen have not been found to date.

The Police Armored Replacement Unit, in addition to training tank mechanics (see above), who belonged to the personnel of the armored car or tank platoons and were also reserve personnel for the crews, also trained specialists for the workshop platoons of the armored companies.

To date only two orders for such training have been found. By order of the Reichsführer-Chief (O-Kdo.In K (1b) 252 E Nr. 58/43) of December 27, 1943, 22 NCO and men were assigned to training of personnel for workshop platoons of armored companies, plus 20 NCO and men by order of the Chief of the Ordnungspolizei (Kdo. In. K (1b) 252 E Nr. 141/44) of October 9, 1944, the latter being assigned to Vienna from January 25 to October 25, 1944.

The assignment to the Police Motor Vehicle School, Police Armored Replacement Unit, applied from the stated dates until the time at which assigning to a workshop platoon took place. The NCOs and men always were to report to the Police Motor Vehicle School, Landstrasser Hauptstrasse 68, in Vienna by 7:00 P.M. on the day before the course began.

They were to be supplied with service clothing for outside service according to the order of July 12, 1941 (MBliV. S. 1303) and its later addenda. In addition, each was to bring along two work suits. Their home bases were to give them one carbine (rifle) with 150 rounds, one 08 or 7.65 mm pistol with 80 or 50 rounds, one 84/98 sidearm and one gas mask.

The assigned personnel were to be provided with travel funds for the trip to the police Motor Vehicle School in Vienna. Their food cards were to be given out at the applicable card offices in exchange for certificates which were to be brought along.

Two Unterführer of the first course were home-garrison capable, and were to be employed exclusively in the armored car workshop.

Each course could form two complete workshop platoons with 20 trained NCO and men.

By the order for the course on October 25, 1944, for the replacement of drivers for the Police Armored Company, ten NCO and men with Class II drivers' licenses were to be assigned by the Police Motor Vehicle School in Iglau, Police Driver Replacement Unit, to the Police Motor Vehicle School, Police Armored Replacement Unit, in Vienna.

Effective 2/5/1945, in the process of releasing action-ready officer candidates from the cadre personnel of the Police Motor Vehicle School in Vienna, three Unterführer were assigned to the Police Armored Replacement Unit, and the transfers of three other Unterführer to the cadre personnel of the Motor Vehicle School were canceled. All Unterführer were to begin service in the Police Armored Replacement Unit on 2/5/ 1945. All of them were to be ready for outside service with police armored companies. The named Unterführer to be used in the Armored Workshop and the motor vehicle echelon until assigned to outside service. They were to be supplied with clothing and equipment according to the Chief's order of 10/24/1944, R Ia (2) 100 Nr. 206/44 (BefBlO. S.383). In addition, each Unterführer was to take along one 08 or 7.65 mm pistol with 80 or 50 rounds, a sidearm, a gas mask and his identification papers.

All Unterführer were to be housed and fed by the Police Armored Replacement Unit. They were to turn in the papers they had brought along at the appropriate card office.

The SS salary books of the Unterführer were to be completed. The Police Armored Replacement Unit was to turn in their driving credentials to headquarters before the NCO were transferred to outside service. At the same time, they were to be instructed to report to the appropriate military office. A report on the successful transfer of the NCO to outside service was to be given to headquarters by the Police Armored Replacement Unit.

Effective February 5, 1945, in the process of measures concerning front-line service in 1945, another eleven Unterführer from the cadre personnel of the Motor Vehicle School were assigned to the Police Armored Replacement Unit, and the assignment of twelve Unterführer to the cadre personnel of the Motor Vehicle School was canceled. All of these officer candidates remained in service at the Motor Vehicle School until further notice. But they were to be kept on file by the Police Armored Replacement Unit for use as personnel ready for action with the armored car workshop and the motor vehicle echelon. Only two of them were to be used as replacements by the Police Armored Replacement Unit.

What courses and formations were made until the war ended cannot be documented at this time for lack of source material. At least the 1st Police Armored Company (see there) was reconstituted again and assigned to outside service as of February 22, 1945.

Off-road driver training with a Hotchkiss tank in a quarry near Purkersdorf. For off-road training the turrets were replaced in order to let the drivers learn to handle the tanks at their full weight. (RS)

What looks like a small downgrade in the picture on page 172 is really a very steep slope that surely made demands on driving ability. Three Hotchkiss tanks are being used here for off-road driver training. (JF)

A former platoon leader who took part in the course for crews of Renault tanks and Steyr armored cars from 6/27/44 to 9/23/44 reports that after the training he returned to his home base and, as of 1/1/1945, saw service with a police battalion in the West. From there he was ordered to the Police Armored Replacement Unit in Vienna for the formation of an armored company. This armored company, set up at the beginning of March 1945, was constituted as follows:

Leader Group with supply train, intelligence and workshop platoons

1st Armored Car Platoon (3 Panhard armored cars)

2nd Tank Platoon (5 Panzer I tanks)

3rd Tank Platoon (5 Hotchkiss tanks)

This armored company was stationed in Vienna-Schwechat. At the beginning of April the company was loaded on a train for action in Yugoslavia. But south of Wiener-Neustadt their transport was halted, since the railroad lines were blockaded by the advancing Russian troops. They were transported back to Vienna, and the company, with other available men, was assigned to the two Danube bridges. Before Vienna was surrounded, the company moved westward, and its remaining forces were finally used to secure the roll ferry at Krems.

The Police Armored Replacement Unit remained in Vienna-Purkersdorf until early April 1945 and was finally to be used in the defense of Vienna. But this did not happen. On April 4, 1945 the unit moved to Schlüsselberg, near Grieskirchen in Upper Austria. All or most of the unit's armored cars were left behind in Purkersdorf, since more of them were not ready for travel. After the war they were still standing at the Gendarmerie barracks and in the surrounding streets.

On May 4, 1945 what remained of the unit was taken prisoner by the Americans at Grieskirchen.

This remarkable twin tank is actually a tank dummy being transported to the firing range, where dummy tanks made of wood and cloth were fired on. Here such a light dummy is being carried atop a Hotchkiss tank. (JF)

Firing training with a Hotchkiss tank. The observer uses field glasses to watch the shots fall on the three targets (dummy tanks) in the background. Firing training with tanks and armored cars was usually done at the training camp at Bruck an der Leitha. (JF)

Personnel of the Police Armored Replacement Unit's replacement comp[any were trained at Vienna-Purkersdorf. The crew of this Panhard armored car wear steel helmets and winter uniforms. Tire chains are used for better off-road performance. (PH)

Workshop personnel were trained at the Police Armored Replacement Unit in mid-1944. In front is a veteran first-series Steyr armored car, behind it a second-series type with ball mantlet for an MG 34. The repairing of damaged vehicles was simultaneously training of workshop personnel for the workshop platoons of the armored companies. (HG)

Armored Training Unit South in Lonigo

After the Italian surrender in September 1943, large supplies of Italian armored vehicles came into Wehrmacht possession. They were used not only to form their own units, but also supplied the police with armored cars.

For the training of all the tank, tank-destroyer and armored scout car personnel using Italian vehicles, the Wehrmacht set up the Armored Training Unit South in Lonigo, near Verona, in October 1943. This unit, under the direction of the Establishment Staff South, likewise founded in October 1943, carried out new establishments and restructurings of German units using Italian equipment.

When the 12th (reinforced) Police Armored Company was restructured in September 1944, the training of its armored personnel and the takeover of Italian tanks took place at the Wehrmacht's Armored Replacement Unit South in Lonigo. Nothing is currently known of the personnel training and armored vehicle use of other police units what were equipped with Italian vehicles.

Training Among The Troops

Naturally, the units constantly worked to improve the training standards of the armored crews, and some courses were also taught among the troops.

The crews of Steyr armored cars used in the Generalgouvernement were largely exchanged for untrained personnel in 1940. The newly arrived personnel were then taught by the units themselves.

From March 10 to 27, 1942, Police Battalion 68 in Amsterdam offered a course for crews of armored vehicles from the Netherlands. Twelve Wachtmeister (SB) of the Schutzpolizei were assigned to this course.

The 9th Police Armored Company (see there) had to train their personnel themselves to use their assigned Russian tanks in 1943, as did the 5th (strengthened) Police Armored Company early in 1944.

Such training by the troop units themselves was carried out very often, as many types of armored vehicles could not be used for training at the schools because they were not available there.

This picture was taken in the Amsterdam area in March 1942. There Police Battalion 68 gave a course for crews of armored cars from the Netherlands. One of the training vehicles was this Wilton Fijenoord tank, which was used by the Ordnungspolizei as "Pzkw. Krupp, holländisch". (BS)

The Armored Platoon of Police Weapons School II in Laon, Belgium

At some Ordnungspolizei schools, armored vehicles were used to instruct personnel in training. The school vehicles were used in antitank defense and instructional classes.

After two French tanks had been assigned to Police Weapons School II in February 1944, three tanks were on hand and the command echelon's armored platoon was set up. The two French tanks were armed with French machine guns. The French tank types, and the type and origin of the third tank, are not known.

Maintenance of the tanks and weapons was up to the K-echelon. Drivers for the two French tanks had to be trained. Crew members (tank gunners and machine gunners) were to be taken, if needed (for training), from the K-echelon personnel. The training of these men had to be done in close cooperation with the 5th (MG) Company or the heavy company.

Because of their high fuel consumption, the tanks were used only for training in close-up tank combat and for instructional presentations of "cooperation with tanks", as well as – to a limited degree, linked when possible with attack training – for the training of tank-destroyer platoons.

The Armored Platoon of Police Weapons School III in The Hague, The Netherlands

Photos prove at least the existence of one tank at Police Weapons School III. Here a British Vickers MK VI battle tank was on hand.

A Vickers MK VI tank with registration number 18246 was used for training at Police Weapons School III in The Hague. This British tank was used there to train gunners for combat against armored vehicles. (BAK)

CHAPTER V
ARMORED SCOUT CAR PLATOONS OF THE GENDARMERIE

General Information

The Gendarmerie provided the police services in the country and in localities of fewer than 2000 inhabitants. It formed Gendarmerie posts with few men and Gendarmerie individual posts, the typical village policemen. In outside (war) service too, the Gendarmerie was used in small commands and posts in the country. Sometimes so-called Gendarmerie Hauptmannschaften were assigned. In some HSSPF there were Gendarmerie action commands.

This "Gendarmerie of individual service" was quite different from the "motorized Gendarmerie", which consisted of units in barracks. The buildup of the motorized Gendarmerie began with an order from the Reichsführer-Chief (O-Kdo O (5) 1 Nr. 46/37) on June 30, 1937. 42 small units (Gendarmerie Readiness) were to be set up, ten being individual Gendarmerie platoons (mot.), each with one officer and 36 men; 18 small Gendarmerie companies (mot.), each made up of two platoons of 1 and 36; 12 Gendarmerie companies (mot.) of three platoons, and two Gendarmerie units of four platoons. The last were stationed at intersections of Reich highways and superhighways, for their chief task was watching traffic on the major roads.

In the course of the war, the motorized Gendarmerie was also built up constantly, and saw service in all combat and occupied areas.

Armored Scout Car Platoon of the 1st Gendarmerie Battalion (mot.)

By order of the Reichsführer-Chief (O-Kdo. I O (4) 50/42) on June 24, 1942, a motorized Gendarmerie battalion, of staff and three companies, was set up for temporary service in the Generalgouvernement and stationed in Warsaw. This motorized battalion was later renamed the 1st Gendarmerie Battalion (mot.). By order of the Reichsführer-Chief (O0Kdo. I O (4) 50 Nr. 41/43) on June 22, 1943, a heavy company was set up for the 1st Battalion in Fraustadt as of July 6, 1943. This heavy (motorized) company consisted of a heavy machine-gun platoon, a heavy grenade-launcher platoon and an armored scout car platoon. The armored scout car platoon was to be established and join the battalion later.

Armored Scout Car Platoons in Other Motorized Gendarmerie Battalions

In June 1943 a 2nd Gendarmerie Battalion (mot.) of four companies was set up in Cannes; it could have had a 4th (heavy) company with an armored scout car platoon, but this cannot be documented. The Chief of the Ordnungspolizei did, though, mention in his directive (Kdo. I O (4) 50 Nr. 62.43) of September 9, 1943, replacements for the motorized Gendarmerie companies, stating that replacements for armored scout car platoons were to be requested directly from the Police Motor Vehicle School (Armored Replacement Unit) in Vienna. Thus there must have been several armored scout car platoons. Nor are there documents that indicate the existence of armored scout car platoons in the 3rd to 6th Gendarmerie Battalions (mot.), which were established later.

Armored Cars in Reserve Gendarmerie Company (mot.) Alpenland 3

The Gendarmerie Company (mot.) Alpenland 3, which saw action in the HSSPF. area of Defense Zone XVIII – headquarters in Veldes – had at least one Italian armored vehicle in September 1943. This vehicle, called Panzer 4 (ital.) (probably an AB 41 scout car), was subordinated by order of the HSSPF. in Defense Zone XVIII, on September 19, 1943, to SS Police Regiment 19 for an undertaking in the Pustjavor-Leskeuz border area.

CHAPTER VI
ARMORED VEHICLE RADIOS

General Information

In 1932 and 1933, attempts were made to equip special Schutzpolizei vehicles with radio sets. These attempts, though, never succeeded to the point where the special vehicles were actually fitted with radios. In 1939 the Test Center for Communications at the Police School of Technology and Traffic in Berlin fitted a Steyr armored car with an Fu G X radio set. This first armored car with a radio is said to have proved itself superbly in many actions, especially in the central sector of the eastern front. This armored scout car laid the cornerstone for radio use in armored cars of the Ordnungspolizei. But until the start of 1943, the Ordnungspolizei's armored vehicles were, as a rule, not equipped with radios. Only one Steyr of the 1941-42 series was fitted with a radio and served as a command vehicle with the Police Armored Car Unit (later 1st (strengthened) Police Armored Company). Messages were carried by motorcycle couriers until early 1943, or sent between armored cars by optical signs (flags, lights).

Only early in 1943 did the use of radios by the Ordnungspolizei increase markedly, as a number of police armored companies were equipped with radios. Experience had shown that radio was an indispensable means of command for the police armored companies. Only by radio was a company leader always able to lead his platoons and also keep in touch with his commander (regimental, battalion or battle group commander). From the company radio site, every radio set in the regiment or battle group could be kept in radio contact. Only through radio connections with all units engaged in combat could successful cooperation be guaranteed. In Russia in particular, where actions often stretched for distances of several hundred kilometers, radio was the only means by which a commander could lead his subordinate units.

As of 1943, all armored vehicles were equipped with transmitting and receiving sets, and were thus very versatile in communications. For example, combat observations could be exchanged among armored cars, or particular events reported. If the platoon leader's vehicle was knocked out, the leader could take command from another vehicle. Besides, an armored platoon might have only two, or even one, vehicle that was fully in touch by radio. The armored platoons could quickly send reconnaissance information or reports to the company leader over distances up to 120 kilometers. Thus countermeasures against observed enemy movements, partisan bands, etc. could be taken at once and, if necessary, reserves could be called in.

Radio contact had also proved to be absolutely necessary for the tank platoons. Only through radio contact was the leader of a tank platoon able to lead his platoon in combat. But only the platoon leader's tank was equipped with a radio transmitter and receiver; the other vehicles had only receivers.

The police armored companies equipped with radios were accompanied by an intelligence platoon. For radio command of his company and contact with other units, the company chief was given a special command radio car with Fu 12 A and Fu 6 radios. The armored platoons generally had command vehicles with Fu 12a radios and two vehicles with Fu 12 radios. In the tank platoons, the platoon leader's tank was fitted with an Fu 5 radio and each other tank with an Fu 2 set.

All armored cars and tanks had to be shielded from interference before radios were installed. Careful shielding was especially necessary in tanks and command radio vehicles, since the ultra-short-wave receivers installed in these vehicles were especially sensitive to disturbances to the generator and ignition system.

The installation of radio sets in armored vehicles was done by radio mechanics of the intelligence platoon of the Police Armored Replacement Unit in Vienna-Purkersdorf.

The main means of communication with a police armored company was the star system. In it, all leaders (company, platoon and tank leaders) communicated on the same wavelength. Communication between the command radio car and another radio site within the regiment or unit was done as line traffic on two wave-lengths.

The Intelligence Platoon

The intelligence platoon of a police armored company was formed as follows:

One intelligence platoon leader (Meister or Hauptwachtmeister),
One radio mechanic,
One equipment maintenance man,
One tank radioman for the company leader's command vehicle,
Three tank radiomen for the first armored scout car platoon,
Three tank radiomen for the second armored scout car platoon,
One tank radioman in reserve.
The tank radiomen of the armored platoons did not belong to the intelligence platoon.

In comparison with a battalion intelligence platoon, the intelligence platoon of a police armored company was considerably weaker in personnel. But in terms of radio equipment it was far superior to a battalion intelligence platoon or even a police intelligence company. The intelligence platoon of a police armored company with two armored car platoons and one tank platoon was equipped with the following transmitters and receivers:

Seven 80-watt Type 80 W.S. a. medium-wave transmitters,
One 20-watt Type 20 W.S. c, ultra-short-wave transmitter,
One 10-watt Type 10 W.S. c. ultra-short-wave transmitter,
Three Type Torn. Empf. b. all-wave receivers,
Four Type Mw. E. c. medium-wave receivers,
Six Type Ukw. E. e. ultra-short-wave receivers.
The equipment of reinforced police armored companies with further armored car or tank platoons was even more extensive.

The Command Radio Car

Every police armored company was assigned a command radio car for the company leader. This command radio car (Kfz. 15 Steyr with four-wheel-drive) was equipped with the following radio sets:

One 80-watt Type 80 W.S. a (Sender a) transmitter with converter unit,
One canister receiver b with transformer unit,
One 80-watt Sender c with converter unit,
One ultra-short-wave Receiver c with converter unit.

To produce electricity for the two transmitters and receivers, the command radio car also had two 12-volt batteries of 105 Ah each. On the right side of the car was a 2-meter staff antenna for the 80-watt transmitter and the ultra-short-wave receiver. On the left side of the car was a Star Antenna d for the 80-watt medium-wave transmitter and the canister receiver. Since only one antenna was present for the transmitter and receiver jointly, radio communication of police armored companies could only be carried on in turn.

In addition, the command radio car carried one standing mast antenna (5-meter mast with Star Antenna a). This antenna could be used only when standing and was set up beside the command radio car.

The crew of a command radio car consisted of the company leader, the intelligence platoon leader, the radio mechanic, one tank radioman and one driver. The company leader could use both speakers and hear receptions from his seat.

The command radio car had the disadvantages of not being armored and having no firepower. Thus it was very sensitive to enemy fire. It was not possible for the company leader to take part in combat with the command radio car.

The Ordnungspolizei would have liked to have armored medium personnel carriers (Sd.Kfz. 251) as command radio cars, since these had sufficient armor, very good off-road capability, and two machine guns. But since this type of vehicle could not be obtained, they had to use the Kfz. 15 Steyr instead, which was not optimally suited for this use.

The Radio Sets

The radio sets used in the armored vehicles will be described in terms of their construction and technical features:

The Fu 12a Radio Set

Consisted of an 80-watt Sender a (80 W.S.a) with a wavelength range of 1120 to 3000 kilohertz for soundless telegraphy and telephoning, a Converter Set 80 a for 12-volt starter batteries, a Canister Receiver b with a wavelength range of 100 to 6970 kilohertz, an Transformer Set E.W.c for 12-volt starter batteries, a 5-meter mast with Star Antenna a and an on-board antenna (Star Antenna d).

The Fu 12 Radio Set

Consisted of an 80-watt Sender a (80 W.S. a) with a wavelength range of 1120 to 3000 kilohertz for soundless telegraphy and telephoning, a Converter Set 80 a for 12-volt starter batteries, a medium-length Receiver c (MwE. c) with a wavelength range of 830 to 3000 Khz, a Converter Set E.U. a for 12-volt starter batteries, and an on-board antenna (Star Antenna d).

The Fu 6 Radio Set

Consisted of a 20-watt Sender c with a wavelength range of 27.2 to 33.3 megahertz (11 to 9 meters) for sound telegraphy and telephoning, a Converter Set 20 a for 12-volt starter batteries, an ultra-short-wave Receiver e (Ukw.E. e) with a wavelength range of 27.2 to 33.3 megahertz (11 to 9 meters), a Converter Set E.U.a for 12-volt starter batteries, and an on-board antenna (2-meter staff antenna for armored vehicles).

The Fu 5 Radio Set

Consisted of a 10-watt Transmitter c (10 W.S. c) with a wavelength range of 27.2 to 33.3 megahertz (11 to 9 meters) for sound telegraphy and telephoning, a Converter Set 10 a for 12-volt starter batteries, an ultra-short-wave Receiver e (Ukw.E. e) with a wavelength range of 27.2 to 33.3 megahertz (11 to 9 meters), a Converter Set E.U. a for 12-volt starter batteries, and an on-board antenna (2-meter staff antenna for armored vehicles).

The Fu 2 Radio Set

Consisted of an ultra-short-wave Receiver e (Ukw.E. e) with a wavelength range of 27.2 to 33.3 megahertz (11 to 9 meters), a converter Set E.U. a for 12-volt starter batteries, and an on-board antenna (2-meter staff antenna for armored vehicles).

With the equipment described here, the following ranges could be reached:

a) From the command radio car to the armored platoons:

Telegraphy when stopped with standing mast antenna: 100-120 km

Telephoning when stopped with standing mast antenna: 40 km

Telegraphy when stopped with on-board antenna: 60 km

Telephoning when stopped with on-board antenna: 20 km

Telegraphy in motion with on-board antenna: 45 km

Telephoning in motion with on-board antenna: 15 km

b) From the command radio car to the tank platoon:

Standing: 4 km

In motion: 2-3 km

c) From the armored platoon leader's vehicle to the armored vehicles:

Ranges as in a). Here radios could be used only with the on-board antennae since the vehicles were not equipped with standing antennas.

d) From the tank platoon leader's vehicle to the tanks:

Ranges as in b).

Radio contact from the command radio car to the tank platoon leader's vehicle, and from the tank platoon leader's vehicle to the tanks, was done only in spoken form. The radio sets in the tanks were operated by the tank leaders, since the Ordnungspolizei's tanks (Renault, Hotchkiss and Panzer I) were usually two-man tanks and had no room for a radioman. Only in the German Panzer III, III and IV tanks was there a radioman in the crew, who operated a machine gun as well as the radio sets and PA system.

Electric Power Production

All transmitters and receivers except the Portable Receiver b were powered by the vehicles' 12-volt starter batteries through converter sets. The Portable Receiver b was run by a transformer set, likewise with power from 12-volt starter batteries. The transformer sets proved themselves considerably better in action than the converters, as they were not so sensitive to dust and moisture and took much less power from the starter batteries. The transformers were also much easier to service than the converters. The converters had to be cleaned and overhauled very often.

Every police armored company was equipped with a Charger Set D to charge the starter batteries of the armored cars and tanks. The power consumption of the converter set for the 80-watt transmitter was 35 amperes. There were also 2.6 amperes of heating power, which was taken off directly from the 12-volt starter batteries. For the converter set of the 10-watt ultra-short-wave transmitter, only 6.7 amperes of power were needed, and 10 amperes for the 20-watt ultra-short-wave transmitter of the command radio car.

With constant use of the 80-watt transmitter, the starter battery would be empty after three hours. For this reason it had to be noted in action that the 80-watt transmitter could be used to transmit only especially urgent and important messages and orders.

The receivers used considerably less power. The transformer of the Portable Receiver b used one ampere and the converter of the medium-range receiver used 4.1 amperes including heating power.

The Speaker System

All armored cars (Panhard and Steyr) and the German tanks (Panzer I, II, III and IV) were fitted with on-board speaker systems. The speaker allowed the crews to communicate, even with the greatest engine and battle noise. The speakers of the Panhard and Steyr armored cars, for example, consisted of a Z 18 speaker box, a Z 33 divider box, and three Z 19 receiver boxes in the Panhard or four Z 19 boxes in the Steyr.

The speaker system of the Panzer I tank, for example, consisted of one Pz 10a box, one Pz 10b box and one sliding-ring transmitter. The speaker systems offered two possibilities. The switch on the Z 18 speaker box could be set in two different ways.

With the "Bord" setting, the vehicle leader could speak to the driver and the gunner via the system (and with the reverse driver of the Steyr). Each crewman had a headset with a microphone and soundproof double earphones. In this case the radioman could take part in talk and was switched onto the transmitter and receiver.

With the "Bord-Funk" setting, the whole crew could listen when the sender switched to "Empfang". When he switched to "Telephonie", the crew could speak to the sender via the speaker system.

The first attempts at fitting armored vehicles with radios were made at the Police School for Technology and Traffic. On May 6, 1940 General Daluege inspected the school in Berlin. At left is a first-series Steyr armored car with an Fu X and frame antenna. (BAK)

The Fu 12a radio set for platoon leaders' armored vehicles, with Star Antenna d. (PFA)

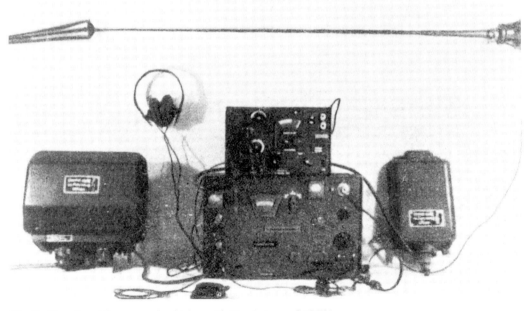

The Fu 12 radio set for armored vehicles, with Star Antenna d. (PFA)

The Fu 6 radio set with 2-meter staff antenna for Sstkw 8 command radio cars. (PFA)

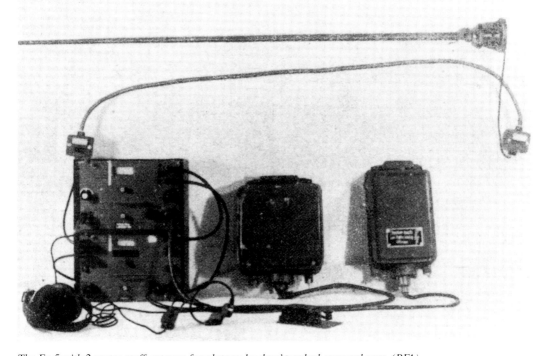

The Fu 5 with 2-meter staff antenna for platoon leaders' tracked armored cars. (PFA)

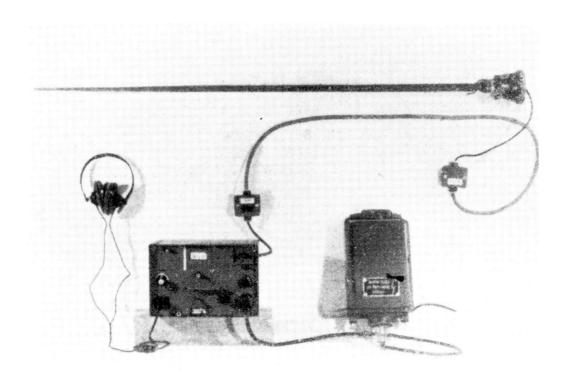

The Fu 2 radio set with 2-meter staff antenna for tracked armored vehicles. (PFA)

One second-series Steyr armored car was fitted with a radio set and frame antenna in 1942. This vehicle, with registration number 20433, was stationed with the Police Armored Vehicle Unit in Krainburg in 1942. (ML)

Another view of Steyr armored car 20433. The frame antenna resembled those that the Wehrmacht used on its six- and eight-wheeled armored scout cars (Sd.Kfz. 232). The Steyr armored car last saw service with the 1st (reinforced) Police Armored Company. (BAK)

During 1943 all Steyr armored cars were fitted with radios. The radio and antenna were mounted at the left, in front of the former machine gunner (now radioman), with the machine gun removed. (PFA)

The Steyr 1500 A/01 Command Radio Car (Sstkw. 8, designated Kfz. 15 by the Wehrmacht) was equipped with an Fu 12a and an Fu 6 radio set. The armored companies were supplied with these vehicles from 1943 on. (PFA)

A Panhard armored car with Fu 12a or Fu 12 radio and Star Antenna d, seen in the yard of the Police Armored Replacement Unit in Vienna in 1943. The installation of radios in armored Police vehicles was done by the intelligence platoon of this unit. (PFA)

The armored car radioman was also the reverse driver of the Panhard armored car.

This tracked Renault tank is fitted with an Fu 5 or Fu 2 radio and 2-meter staff antenna for armored vehicles. The Renaults had the antenna mounted at the left rear of the vehicle. (PFA)

The radio antenna was mounted in a special bracket on the track apron at the right rear of the Hotchkiss tank. This mounting position was also chosen by the Wehrmacht. (PFA)

The German armored vehicles taken over from the Wehrmacht by the Ordnungspolizei normally had no changes made to their radio equipment. Here Panzer IV and II tanks of the 13th (reinforced) Police Armored Company are seen. (BAK)

CHAPTER VII
PAINTING THE ARMORED AND OTHER VEHICLES

In June 1936, the first orders for the painting of vehicles, weapons and technical equipment of the Ordnungspolizei were published in RMBliV. on page 897 of Rd.Erl. d. RuPrMdI v. 3.7.1936 – III S I de 12 Nr. 28/36 K. Accordingly, dark green, Color Tone No. 30 on the color chart for vehicle paints (Nr. 840 B 2) of the Reich Branch for Delivery Conditions (RAL), was to be used to paint motor vehicles and other vehicles of the Police – bicycles excepted. For vehicles that served special purposes, this color could be disregarded. The same color was also to be used for intelligence and weapon-technical equipment, as well as lights, as long as the latter needed to be painted. If any vehicles and the noted equipment were already painted in colors that varied from this color, a change of color was to be carried out only when new paint became necessary for them.

By 1938 the special police vehicles of the Ordnungspolizei had not been given a new coat of paint. A directive from the Reichsführer-Chief in the Ministry of the Interior, dated January 20, 1938 (O-Kdo. T (2) 205 Nr. 2/38) to the police offices at which ready-to-drive special police vehicles (Pskw.) were on hand specified that for parade purposes the vehicles listed in the appendix were to be put into good driveable condition by March 1, 1938. The special vehicles were also to be painted uniformly in Color Tone 30, semi-matte, as on the color chart for vehicle paint Nr. 840 B 2 on the list of the Reich Branch for Delivery Conditions (RAL), to be ordered from the Beuth firm at Dresdener Strasse 97, Berlin SW 14, or the firm of Otto Hieronymi in Göttingen.

The old paint was to be left on the vehicles. As new paint, the artificial lacquer "Gerrolux" (Tone 33 semi-matte RAL 30 (Police Green), obtained from the "Ilag" firm Industrielackwerke A.G. in Düsseldorf-Gerresheim, guaranteed to hold on the old paint, was to be used. One kilogram of this paint covered some 8 to 10 square meters. Whether the armored cars and tanks were also painted RAL 30 "Police Green" is not definite.

This Daimler special vehicle of the Prussian Schutzpolizei, sprayed in a very light color, was presumably, if it was still driveable and usable, painted in "Police Green" RAL 30 paint according to the order of January 20, 1938. (ER)

In any case, a change in the regulations for painting Ordnungspolizei vehicles and equipment was made in August 1942 and published in MBliV. S. 1673 (Rd.Erl. d. RFSSuChdDtPol. im RMdI v. 14/8/1942 – O-Kdo I K (3) 201 Nr. 7/42). To save material and manpower, all new Ordnungspolizei motor vehicles were delivered, effective immediately, in Wehrmacht gray paint. For the same reason, all Ordnungspolizei motor vehicles, including trailers, that needed new paint were to be painted uniformly (including fenders) in matte gray paint, using Color Tone 46 RAL 840 B 2. The same applied to any other vehicle equipment. The previous police green color could be obtained and used only for touching up. The order of 7/3/1936 (MBliV. p. 897) was replaced by the new order insofar as it dealt with the painting of motor vehicles.

In October 1942 a hint of a change in the color tone designations of paint for Ordnungspolizei motor vehicles and equipment appeared in the publication of MBliV. (p. 1972), with an order from the Reichsführer-Chief on October 5, 1942 (O-Kdo I K (3) 201 Nr. 7 II/1942). Completing the order of August 14, 1942, it stated that the Color Tone 46 RAL 840 B 2 shown in Section 2 meanwhile had been given the new designation "RAL 7021".

In 1943 the Ordnungspolizei followed the Wehrmacht's change in paint colors. In an order of the Reichsführer-Chief on April 7, 1943 (O-Kdo I K (1c) 201 Nr. 26/43), published in April 1943 in MBliV. on page 590, it was said of the painting of Ordnungspolizei vehicles that newly built vehicles would not be delivered to their bases with gray paint any more, but with dark yellow. When it was necessary to repaint vehicles already on hand, the dark yellow paint as used by the Wehrmacht was to be used whenever possible (the RAL color tone would be announced). Remaining supplies of gray paint could be applied for touching up. As far as the use of the vehicles allowed, they were to be camouflaged according to local conditions by the individual bases or formations by using gasoline-soluble camouflage paste (olive green, red-brown or dark yellow). If the colors of canvas truck covers contrasted too strongly with the body color, the covers were likewise to be colored with camouflage paste. Winter camouflage in snowy areas consisted as before of white paint with emulsion coloring. The order of 8/14/1942 (MBliV. p. 1673) was inapplicable when existing conditions opposed it.

Further directives for painting and coloring Ordnungspolizei0 motor vehicles and armored cars have not been found. On the basis of photos, though, it can be seen that the Ordnungspolizei used the same colors as the Wehrmacht during the war, which also allowed new armored vehicles to be assigned via the Wehrmacht.

Whether the tanks and armored cars of the Police were sprayed "Police Green" (RAL 30) or "Wehrmacht Gray" (RAL 840 B2) until August 1942 or both colors were used simultaneously cannot be told from black-and-white photos. These Steyr armored cars in the yard of the Armored School in Purkersdorf were painted a dark color in the spring of 1942. (RS)

This Steyr armored car has been overhauled and painted the new dark yellow color on being assigned to the newly formed 2nd Police Armored Company. The insides of the doors, like the sprayed No. 30 chassis, have also been painted dark yellow. (AW)

In winter use the Ordnungspolizei's armored vehicles were covered with white emulsion paint that could be washed off, like this Steyr armored car of the 4th Police Armored Company. (BAK)

These two vehicles of the Bozen Police Regiment, an Italian AB41 armored car and a Ford Stkw. 14, have been painted dark yellow and then camouflaged with olive green and red-brown camouflage paste. (RE)

In the Gorizia area this T34 tank has been given new camouflage paint with basic dark yellow plus shadings with camouflage paste. At first the German crosses were newly painted in plain white. The black bars followed in the second process. At the left front of the bow plate is the camouflage headlight (night marching device) for driving at night. (JF)

CHAPTER VIII
MARKS OF RECOGNITION AND INSCRIPTIONS

General Information

The armored vehicles of the Ordnungspolizei could bear the following markings and letterings: German emblem, German crosses, official police emblems, Ministry of the Interior numbers, chassis numbers, troop emblems and further emblems and letterings.

These identifying marks and inscriptions could be applied to the vehicles together or in varying combinations.

The German Emblem

The national emblem for service vehicles of the Ordnungspolizei and Fire Police was illustrated in the January 1938 issue of the RMBliV,m on page 16 of the order of the Reichsführer-Chief dated 12/27/1937 (O-Kdo T (2) 201 Nr. 37/37). Accordingly, all service vehicles of the Schutzpolizei of the Reich and communities, the Gendarmerie (including motorized units) and the Fire Police, with the exception of production cars, Pskw., motorcycles and normal ambulances of the Fire Police, were to bear the national police emblem. The emblem, in the form of an adhesive picture, was to be applied to both sides of the vehicle in readily visible places, especially on the drivers' doors. The greatest width of the emblem, measured between the eagle's wing tips, was 20 cm. The eagle's head always had to face forward, so that for every vehicle two different labels had to be made. For protection against weather conditions, the emblems were to be painted over with an acid-free colorless paint. As before, newly made vehicles were delivered without the national emblem, which had to be applied by the units after delivery.

The emblems described here were also applied to the turrets of the armored cars, and later to the turrets of the tanks, of the Ordnungspolizei. As stated in the order, these emblems could not be affixed to special police vehicles until then.

From March 1943 on, the emblems were not applied, as was stipulated on page 482 of the RMBliV of 1943, in an order from the Reichsführer-Chief dated 3/19/1943 (O-Kdo I F (2b) 210 Nr. 49 III/43) concerning national emblems on service vehicles of the Ordnungspolizei. To save raw materials and work time, the instructions on lettering fire trucks in the "regulations about the building of fire vehicles" were abolished. In the future, the vehicles were delivered by the manufacturers without police emblems or lettering. The Ordnungspolizei's other service vehicles were not to have national emblems applied either. Labels on hand could still be used. The order of 12/27/1937 (MBliV. 1938 p. 16) was abolished.

German Crosses

No orders about applying German crosses to armored police vehicles have been found. Here, though, the Ordnungspolizei followed the Army's practices. It is known that when the Polish campaign began, the Steyr armored cars bore only the national emblem. During and after this campaign, white crosses, like those on armored Army vehicles, were painted on the armored cars. The Army High Command had already ordered on 7/13/1939 (HM 39, Nr. 525) that a white bar cross should be painted on all four sides of every armored vehicle. The white bar crosses were painted on the front and rear bodies, both sides of the turrets and the turret roofs of the Ordnungspolizci's Steyr armored cars.

To change the order of 7/13/1939, the Army High Command ordered on 10/26/1939 that the white bar crosses were to be removed from the fronts of all armored vehicles. Those on the sides and rear were to be altered into open crosses, like the recognition symbols on German airplanes. The width of the bars should be about 2.5 centimeters. The crosses were not to be painted on the turrets as before, but on the upper bodies.

A Steyr armored car (registration number 20441) of the Police Armored Car Unit Center in mid- 1942, bearing the national (eagle facing forward) emblem, bar crosses and registration number. (BAK)

The further marking of the Ordnungspolizei's armored vehicles with German crosses deviated from this order. Only at the beginning of the Russian campaign were armored police vehicles again marked with bar crosses, open crosses like those of the Army. Probably for better and quicker identification by German troops (who had no Steyr, Tatra or 7 TP armored vehicles and thus were not familiar with these vehicles), the white crosses were made wider than "normal", or the black inner crosses were made thinner. Some crosses had only a horizontal black bar, with the vertical bar all white. All-white crosses were still used in Russia until 1942 as well. In the further course of the Russian campaign, the bar crosses usually resembled those of the Army, especially on the Panhard, Renault and Hotchkiss vehicle types also used by the Army, as well as German and, later, Italian tanks. It is interesting that almost all of the Ordnungspolizei's tanks had a bar cross on the bow, and some also on the turret top. This was important for recognition by German troops, but the markings made excellent targets for enemy antitank defenses, and were thus rarely used by the Army. On captured Russian tanks the bar crosses, as on captured tanks used by the Army in general, the crosses were painted oversize, again to make recognition by German troops easier. German crosses were also painted on the turret tops of captured Russian tanks.

Official Markings

License plates with the letters "Pol" followed by a number were issued centrally according to orders of 7/10 and 7/15/1935 (MBliV. p. 882, 910), 6/23 and 10/1/1937 (MBliV. p. 1034, 1627), 11/4/1940 (RMBliV. p. 2069), 4/21/1941 (RMBliV. p. 767) and further additions. The police number groups were given out to the Reich governors, highest state authorities, Prussian government presidency and Berlin police headquarters. These offices were to handle the further distribution within their areas. They had to be sure that only closed groups of numbers were given out. Thus the number indicated the home base of the vehicle. During the war, vehicles of new units were sometimes sent by various police offices in the Reich, so that it was no longer possible to tell the unit's home base by the license numbers. The vehicles of the armored police companies that were established in Vienna or supplied by the Armored Equipment Office usually used license numbers from 143001 to 145000, "Reich Governor in Vienna".

As a rule, armored vehicles were not given police markings. Unlike the Wehrmacht, which gave its wheeled and halftracked armored vehicles WH numbers, the wheeled and halftracked armored vehicles of the Police were not given such numbers.

Only a few deviations from this rule could be found: A Büssing armored car with railroad wheels, used by Rail Armored Platoon 3 of the Army as an armored railcar in 1939, bore the number Pol-10019, a number assigned to the Munich police headquarters.

In SS Police Regiment 18 in Greece, all L 6 armored cars and assault guns bore police numbers of unknown origin (Pol-228...), such as an L 6 tank with Pol-228494.

A Daimler DZVR special police vehicle of the reinforced Police Armored Platoon of Berlin bore a police number from the numbers of the Berlin police headquarters (Pol-2672) in 1945.

Ministry of the Interior Numbers

The Prussian Ministry of the Interior had already begun in the early 1920s to introduce its own ministry registration numbers in addition to the usual motor vehicle numbers, such as chassis, engine and body serial numbers. This number was on a special plate, which was to be attached to the vehicle. Along with the number of the vehicle, for example 436, the year of manufacture was also given as of October 1924, so that a 1924 vehicle might have the ministry number of 436/24.

After the Reich was set up, this system was taken over by the whole police and the Ministry of the Interior's numbers became Reich Ministry numbers, whereby all former Prussian numbers were taken over and all vehicles from other provinces were given subsequent numbers. For example, special vehicles and other vehicles made in Saxony in the 1920s always had much higher numbers than Prussian vehicles of the same age.

On 5/23/1940 an order from the Reichsführer-Chief in the Ministry (RMBliV. p. 989) assigned Reich Ministry numbers to Ordnungspolizei vehicles. Here the Reichsführer-Chief specified that Ordnungspolizei vehicles that did not yet have registration numbers were to be given numbers at any time by special order, and the number plates were to be provided for attachment to the vehicles at the same time.

The plates were to be attached as follows:

- If possible, to the dashboard of the vehicle, otherwise to the left door frame.
- On the front fender of the vehicle.
- To the upper body panel of a sidecar behind the seat back.

Since many vehicles saw long-term service away from their home base, the plates that were sent out were to be preserved carefully and attached to the vehicles on their return to their base. In the event that a vehicle in service away from its base did not return, but was assigned to another base, then the plate for the vehicle was to be sent to the new base with the necessary explanation. The new base was then responsible for the proper attachment of the plate. The Reichsführer-Chief pointed out that the Ministry number plates were only made in one series and represented the only way for him to keep track of the police vehicles. At the same time, he reminded that all reports about Ordnungspolizei vehicles should, under all conditions, should include the Ministry number along with other information.

By order of the Reichsführer-Chief in the Ministry (O-Kdo I K (4) 201 Nr. 8/41), motor vehicle papers and Ministry numbers for captured and confiscated Ordnungspolizei vehicles were regulated. Now Ordnungspolizei bases that possessed captured or confiscated vehicles and trailers made outside Germany, and which did not yet have authorization for use in the Reichs, should be reported to the appropriate office according to Section 21 of the StVZO. In addition, the Reichsführer-Chief was to be given the following data on all confiscated, purchased or captured motor vehicles (including German ones) that had not yet been assigned a Ministry number: 1. Manufacturer, 2. Type of vehicle, 3. Type of body, 4. Type designation, 5. Number of seats, 6. Engine displacement or power, 7. Chasis number, 8. Engine number, 9. Year built, 10. Odometer reading, 11. Condition, 12. Home base, 13. Remarks (here it was to be stated whether, and if so in what report, the issuing of a Ministry number had already been requested).

It soon became apparent that the factory serial numbers of the captured vehicles and trailers could not always be determined. According to an order from the Reichsführer-Chief on 11/1/1943 (O-Kdo I K (1c) 202 Nr. 342 II/43), in such cases the Ministry number assigned to the vehicle was also to serve as the chassis number. So that it could always be determined that the number had been assigned by the Police, the letter "P" was added before the number. The number imprinted on the Ministry plate to indicate the year of manufacture was then omitted. The number should be placed, easily legible, at the front of the chassis frame and on any factory plates on the vehicle.

From 1942 on, Ministry numbers also appeared on the outsides of armored vehicles, but no order for this has been found. The attachment of Ministry numbers was probably done along with the centralized training and establishment of armored units in Vienna. The Ministry numbers, already five digits long by then, were painted on the front and rear ends of Steyr armored cars in white paint on a number plate, with the first three digits always smaller than the last two. On the Panhard tanks the numbers were painted directly on the front and rear of the body. The Ministry numbers were painted directly on tanks as well, but usually only on the front. Hotchkiss tanks, for example, had them at the right side near the front, as did Renault tanks, with very large numbers. The size of the painted-on numbers varied, but the last two digits were always larger and served as a "short recognition form" for the vehicle.

Troop Markings

In order to make known that a vehicle belonged to a particular unit, the Army and Waffen-SS used a variety of markings, which were applied to all vehicles. The type and application of these symbols was partially regulated by orders.

Although attaching official markings to armored Ordnungspolizei vehicles was not customary, there were some exceptions. This L6 tank of SS Police Regiment 18 bears number Pol-228494 in Athens in May 1944. (BAK)

Two Renault tanks of the Police Armored Unit are seen in Krainburg in the summer of 1942. The Ministry number was painted on the front of the right side of this type of tank, with the last two numbers painted bigger to identify the tank. The tanks also bore national emblems and German crosses, plus a troop emblem (Edelweiss) on the right track apron. (ML)

No orders for the Ordnungspolizei on this subject have been found. But the system of troop emblems was also used in part by the Ordnungspolizei's armored troops, though only to a limited degree. For example, troop emblems have been found for the armored unit of the Police Mountain Regiment (Edelweiss flower), Police Armored Company 2, formed in 1943 (diamond with number 2), the reinforced Police Armored Company 12 (coat of arms), and reinforced Police Armored Company 13 (turret). The armored platoon that served in Danzig in 1939 also had its own troop emblem, a skull with an SS rune above it.

Other Emblems and Lettering

At the beginning of the Russian campaign, white swastikas were painted on the turret tops of Steyr armored cars for identification of German vehicles from the air. Large swastikas were used particularly in the Police Regiment Center's zone, where big ones were painted on the sides of the Steyr armored cars, the radiator grills of the tanks from Holland, and the bow plates of the 7TP tanks. As in the Army, air recognition cloths were used, red rectangular cloths that showed a black swastika in a white circle. These could be spread out on the vehicle and removed again quickly. Chassis numbers were marked on the type panel inside armored Wehrmacht vehicles, but also struck into the outside of the armored hull or body. But these hammered-in numbers were not easy to recognize, especially after several coats of paint. Thus it was often the custom to paint the chassis number on the outside of the vehicle, usually in the bow area, for clear identification.

In the Ordnungspolizei, painted numbers have been confirmed to date only on the Steyr armored cars, There the number was painted inside on all four upper parts of the side doors. When a door was opened, the number was visible from outside.

Whether chassis numbers were also painted on the insides of doors and hatches of other armored vehicles could not be confirmed for lack of photographic evidence.

The Army's armored vehicles, as a rule, bore turret or vehicle numbers that showed the organization of the vehicle in its unit. A widespread system of three digits was used, in which, for example, 124 stood for the fourth vehicle in the second platoon of the first company. Further systems could concern the color of the numbers or unit and tactical symbols.

In the Ordnungspolizei, turret numbers were fairly rare, since the Ministry number was obviously used as the most important identifying mark. A uniform system of issuing turret numbers could not be confirmed.

In the 2nd Police Armored Company, established anew in mid-1943, the 3rd Tank Platoon had a turret number system. From the five-digit Ministry number that was borne on the front armor, the last two (large) digits were repeated on the rear, and the last digit was also painted in large size on the back of the turret.

In the 8th Police Armored Company too, the last two digits of the Ministry number were painted in large size on the turrets of the Panhard armored vehicles as turret numbers.

In the 5th strengthened Police Armored Company, the T-34 tanks in the tank platoons are known to have had three-digit turret numbers, such as 210, but they may have come from previous owners of the tanks (probably a Wehrmacht unit). The same is true of the three-digit turret numbers (626, 727) on the Panzer IV tanks of the 13th strengthened Police Armored Company. The 10th Police Armored Company's Panhard tanks also bore a one-digit number on their bodies that had nothing to do with their Ministry numbers and were apparently a thorough numbering of the tanks. For example, the Panhard with Ministry number 27187 bore number 4. The only number that fit into the Army's numbering system was found on an assault gun on Armored Platoon 18 in SS Police Regiment 18. This vehicle, numbered 184, could have been the fourth assault gun of the 18th platoon.

Vehicle names had nothing to do with central regulations, but were individual names attached or painted on by unit leaders or crews.

In the armored vehicle platoon used in Danzig in 1939, both Daimler and both Steyr armored cars bore names (Memel, Saar, Ostland and Sudetenland) painted on in large white letters. Similar names were borne by three Steyrs of the armored platoon of Police Regiment Center when the Russian campaign began in 1941. The vehicles bore the names Danzig, Memelland and Pommern painted on their sides in large white letters. Two vehicles from Holland in Police Regiment Center bore the names "Arnhem" and "Den Haag".

This Steyr from an armored platoon of Police Regiment Center has the national emblem on its turret and a large swastika and the name "Memelland" on its sides. The photography van in the foreground bears the number Pol-36105 (first issued to the Reich Protectors in Bohemia and Moravia) and a dog's head as a troop emblem. (PH)

These two vehicles of the 13th (strengthened) Police Armored Company, seen on the Atlantic coast of France in the summer of 1943, could serve as models of how to affix markings. The Steyr has a bar cross, Ministry number 22771, and the troop emblem on its rear, and the Panzer IV has turret number 727, a bar cross, Ministry number 18140 (at right above the cross) and troop emblem. (BAK)

A Russian vehicle of the 13th strengthened Company of Police Regiment 14 bore the name "Emil" in large white letters. This was the commander's first name.

In the 12th Police Armored Company, and M 15 tanks had the name "Gretl" on the right side of its body, and in the 13th strengthened Police Armored Company there was a flame throwing tank with the name "Greta" on its right front. These could have been names of wives or girl friends of commanders or crewmen.

Men of the 13th (strengthened) Police Armored Company show French girls their Steyr armored car in the summer of 1943. On the insides of the doors the chassis number of the vehicle can be seen, with a bar cross in the center of the turret side. (BAK)

CHAPTER IX
UNIFORMS AND EQUIPMENT

General Information

Within this study we cannot deal with all the pieces of police armored troops' uniforms, but will rather offer an overview and portray special equipment and uniforms of the armored troops. The crews of the state Schutzpolizei's special vehicles were issued varying protective clothing. As a rule, these consisted of leather suits, similar to those of drivers, or combination suits of cloth with asbestos fibers. While in service with the special vehicles, they wore protective leather helmets. Goggles were also widespread. No clothing regulations have been found.

After the introduction of the new armored cars in the Ordnungspolizei, the use of leather helmets, with and without peaks, can be documented, but so can the further use of the old Austrian M 35 helmet with neck protection. All leather helmets had the large national emblem of the police on the front.

Otherwise the crews of the armored vehicles were issued the same special clothing as the drivers: leather hood, goggles and long coat.

The introduction date of the combination suit for crews of armored vehicles is not known; the first uniform regulations seem to have been made in mid-1941. Along with regulation issues of uniforms and equipment, of course, captured items were also used if they proved to be comfortable or useful.

Clothing of Police Officials in Outside Service

In an order from the Reichsführer-Chief on 7/12/1941 (O-Kdo II W 1.10. Nr. 98/41), Appendix 1, PBklV. Part II, equipping officials in the police battalions was specified. They received the following pieces of uniforms and equipment:

1. House cap (1)
2. Weapon or field blouse (2)
3. Breeches (1)
4. Long trousers (1)
5. Coat (1)
6. Neckerchief (2)
7. Denim suit (1)
8. Woolen gloves, (1 pair)
9. Earmuffs or head protector (1)
10. Boots (1 pair)
11. Laced shoes (1 pair)
12. Leggings (1 pair)
13. Under jacket (1)
14. Undershirt (3)
15. Under shorts (3)
16. Socks (3 pairs)
17. Steel helmet (1)
18. Body belt with pouch and lock (1)
20. Handcuffs (1)
21. Bullet pouch (1)
22. Food bag with strap (1)
23. Field pack (1)
24. Coat belt (3)
25. Drink bottle and cup (1)
26. Cooking utensils (1)
27. Cooking utensil belt (2)
28. Tent canvas with accessories (1)
29. Drill shoes (1 pair)
30. Drill shirt (1)
31. Running pants (1)
32. Swimming trucks (1)
33. Training suit (1)
34. Closing chain with lock (1 per 10 men)
35. Map case (1 per 10 men)
36. Signal horn (1 per 30 men)

All officials were to bring along on duty, from their own supplies: eating utensils, sewing, cleaning, shaving and washing articles.

Police medical officers wore a red cross armband on duty and carried two medical kits and a bottle in place of the bullet pouch and the drink bottle and cup.

For sentries and guards, a pair of mittens, a pair of felt overshoes, and a fur coat were issued when needed.

Only drivers were also issued a rubber coat, a work suit, a pair of fur-lined leather gloves, a fur coat and goggles.

Every official going into action was to be issued, from police supplies, two camp blankets and one hand towel, plus about ten percent as supplies.

Special Clothing for crews of Special or Armored Vehicles

According to an order from the Reichsführer-Chief on July 12, 1941, Section M, the following special clothing was issued, in addition to their basic supplies (see above), for crews of special vehicles: a cloth combination suit, a combination suit of rough gray denim, and a crash helmet without a peak.

With the Reichsführer-Chief's order of 12/15/1942 (O-Kdo II W 2 100 Nr. 401/42), Section M of the order of 7/12/1941 (MBliV. p. 1303) was expanded. Accordingly, from then on the following special clothing was issued to the crews of special vehicles:

No.	Designation of the object	Authorized
1	2	3
1	Black field cap (fore-and-aft)	1
2	Black cloth field jacket	1
3	Black field jacket of denim	1
4	Black field trousers of cloth	1
5	Black field trousers of denim	1
6	Short leggings, pair	1
K7	Tricot shirt with blue-gray chest panel (replaces usual issue of undershirts)	3
8	Gray-blue collar	3
9	Long black tie	1
10	Blue work suit	1
11	Goggles	1

Items 1 through 6, 8 and 9 could only be work on service with armored vehicles. The items listed with a "K" were given with a debit for clothing money. The police headquarters (clothing delivery places) that were to set up armored units or train armored car crews requested the lacking pieces of special clothing from the Chief of the Ordnungspolizei. As long as the armored car crews of the units in outside service were not yet supplied with black suits, leggings, shirts with collars and long ties, the units were to request what they needed (including 10 percent as supplies) from the appropriate replacement supply places; they in turn requested what they needed from the Chief of the Ordnungspolizei.

The pieces of service clothing formerly issued to the armored car crews and no longer to be used were to be turned in. Of them, the green cloth combination suits and the helmets were to be turned it to the appropriate replacement supply places (police supply stations), and the other items added to the unit's supply of clothing. The available denim combination suits were to be used as work suits. Steel helmets were to replace the formerly issued helmets. Needed steel helmets were to be requested along with pieces of special clothing.

In the "Special Instructions for Supplying No. 1" given out by the Chief of the Ordnungspolizei (Kdo. I-Ia (Ib) on January 5, 1943, an extract from the Reichsführer-Chief's order of 12/15/1942 (O-Kdo II W 2 100 Nr. 401/42) from MBliV. p. 2338 was printed under the heading "Special Clothing for Crews of Police Armored Vehicles". The complete text of Section B :Clothing Supplying of Police Units" (see above) appeared there.

No.	Designation of the object	Number per man
1	2	3
1	Black field cap (fore-and-aft)	1
2	Steel helmet	1
3	Black cloth field jacket	1
4	Black denim field jacket	1
5	Black cloth field trousers	1
6	Black denim field trousers	1
7	Green coat	1
8	Work suit	1
9	Laced shoes, pair	2
10	Short leggings, pair	1
11	Woolen gloves, pair	1
12	Under jacket	1
13	Head protector	1

No.	Designation of the object	Number per man
1	2	3
14	Tricot shirt with blue-gray chest panel	3
15	Gray-blue collar	3
16	Black tie	1
17	Under shorts	3
18	Socks, pair	3
19	Body belt with pouch and lock	1
20	Signal pipe (only for leaders and subleaders	1
21	Food bag with strap	1
22	Drink bottle with cup	1
23	Cooking utensils	1
24	Canister	1
25	Coat belt	3
26	Cooking utensil strap	2
27	Tent canvas with accessories	1
28	Goggles	1

For service, the following were to be taken along by the police units as spares: of numbers 1 and 3 to 18, ten percent, and of numbers 2 and 19 to 28, five percent.

Cold-weather clothes were delivered from the appropriate replacement supply depots by special order of the Chief of the Ordnungspolizei.

On June 21, 1943 the Reichsführer-Chief's order (O-Kdo I W 2a 100 Nr. 162/43) brought another change in Appendix 1, Section M, of PBklV. Part II, which was published on page 1050 of the BMliV. on June 30, 1943. Accordingly, the following additions were to be made by hand to the plan given in the order of 12/15/1942 (MBliV. p. 2338) for Appendix 1, Section M, PBklV. Part II and in Appendix 1a of the specified order as no. 12 or 29:

In Sp. 2: "Protective suit (only for crews of flame throwing tanks)".

In Sp. 3: "1".

The distribution of the required protective suits took place by procurement or special request.

It is interesting that in this order, both the 11-point Appendix 1a of 12/15/1942 and the 28-point Appendix 1a of January 5, 1943 were referred to.

Training on a Steyr armored car, probably in 1938 or 1939; as yet there is no national emblem on the vehicle. The men wear leather hoods and fore-and-aft caps. The dark coats, probably blue, are without insignia, the scarves are of private origin. (MF)

Armored crews in the Russia-Center zone are being given orders in the summer of 1941. The soldiers wear cloth combination suits with dark brown collars and cuffs. The goggles vary. (BAK)

An armored car crew of the Armored Car Unit of Vienna, seen in Prague in June 1942. The soldiers wear rough gray denim combination suits and peaked leather helmets, which were actually issued only to motorcyclists. (AW)

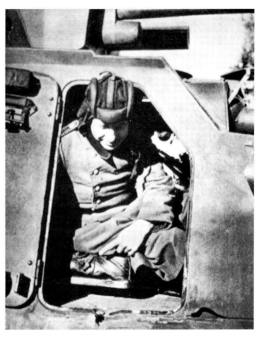

Crewmen of a Steyr armored car with Police Regiment North in the summer of 1941 wear black unpeaked helmets and cloth combination suits without cuffs.

Captured pieces of equipment like this Russian leather helmet were used by armored crews, such as this crewman of a vehicle from Holland in Police Regiment Center in the summer of 1941.

By order of 12/15/1942, the issuing of black special clothing was announced. Leather helmets were to be turned in and replaced by steel helmets. These men of an armored platoon of the Police Armored Unit, seen at Garmisch in the autumn of 1942, wear black helmets with their new black uniforms. (WP)

As instructed on 11/12/1934, the Army introduced a black special uniform for the armored troops. Here an NCO of the Wehrmacht wears the black field jacket and black field cap of the Army. On the right breast is the national emblem, on the left the ribbon for Iron Cross II and the silver armored combat emblem. (JF)

As ordered on 12/15/1942, the Ordnungspolizei also introduced black special clothing for their armored troops. Here a Revier-Oberwachtmeister of the Schutzpolizei wears a black field jacket and field cap. The national emblem is sewn to the left upper sleeve of the jacket; on the left breast is an order bar. (JF)

A Revier-Oberwachtmeister of the Schutzpolizei wears the black special clothing for armored crews of the Ordnungspolizei, consisting of a black fore-and-aft field cap, black cloth field jacket, black cloth field trousers and laced shoes. The policeman also wears gloves, a body belt with clasp and pistol holster, and (not according to prescription) a white shirt and black tie. (JF)

Members of Workshop Platoon 14 of the 5th reinforced Police Armored company show a wide variety of uniforms as they work on a T34 tank in the summer of 1944: Berets, fore-and-aft caps and field caps of varied materials, blue work suits, denim combination suits, shorts and long trousers. (JF)

Members of SS Police Regiment Bozen wear summer or tropic uniforms in the summer of 1944. The commander of the tank, at left, has his jacket tucked into his trousers. (RE)

Various uniforms are worn by members of the 2nd Police Armored Company in October 1944: Black armored uniforms, work suits, draped tent canvases, cloth coats and motorcyclists' rubber coats. (AW)

Fairly uniform winter clothing is worn by members of the 2nd Police Armored Company in February 1944. Almost all the soldiers wear the jacket of the reversible winter suit (one side white, one gray) plus leather and felt boots. Their trousers and headgear vary. (AW)

CHAPTER X
THE ARMORED VEHICLES OF THE ORDNUNGSPOLIZEI

Procurement and Supplying of Armored Vehicles

Until the war began, in fact until November 1939, the types and numbers of armored vehicles used by the Ordnungspolizei can be kept track of quite exactly. It can be determined that at that time the Ordnungspolizei had about 43 special police vehicles (Pskw.) and 37 newer-type armored cars (Pzkw.): 24 Steyr, 10 Tatra and 3 Skoda. The special vehicles had been taken over from the police forces of the states, and the armored cars could be taken over from the police (or army) of Austria or Czechoslovakia. After the Polish campaign ended, the types and their numbers can be traced only somewhat precisely, since no Ordnungspolizei statistics of the armored vehicles they used after November 1939 have been found to date.

Between the Polish and Russian campaigns, the Ordnungspolizei took over two more Steyr armored cars, an unknown number of armored scout cars from Holland, and an unknown number of Polish 7 TP armored cars.

Special production of armored vehicles for the Ordnungspolizei took place only in 1941, when the SS ordered additional ADGZ armored scout cars from the Steyr firm for the police, and 1942, when Steyr armored cars were taken over. In the contemporary police training materials there is very little information on the armored vehicles that were used. We write here only of police armored vehicles in general, and technical data are given only for the Steyr armored car and depict only the Steyr and the Renault tank. In the periodical "Die deutsche Polizei" photos and very short reports on the armored cars or armored troops appeared now and then, but only for propaganda purposes.

All the armored vehicles assigned to and used in the Russian campaign had to come from Wehrmacht production or materials captured by the Army. It is interesting that in Wehrmacht records only a few references to captured armored vehicles being turned over to the Ordnungspolizei can be found.

Thus it cannot be determined exactly how many German or foreign armored vehicles saw service with the Ordnungspolizei. Only a few sources offer even fragmentary information.

In an order from the Reichsführer SS and Chief of the German Police on September 15, 1941 (O-Kdo I K (2) 252 Nr. 79/41) it was then announced that a number of new armored cars had just been delivered to the Ordnungspolizei. Thus it was necessary to begin training drivers for these vehicles at once. Two months later this order was revised or changed by an order from the Reichsführer-Chief on 11/7/1941 (O-Kdo I K (2) 251 Nr. 186/41) on the formation of armored units and training of crews. Now it was stated:

"Within the Ordnungspolizei, armored units will be established first. For their use there will be armored scout cars (wheeled vehicles) and tanks (tracked vehicles). The crews of the armored scout cars (Steyr) consist of one leader, four machine gunners and two drivers each. In all, 35 crews will be formed, including relief crews.

The tanks are captured French vehicles (Hotchkiss and Renault), which are to be crewed by one leader, who is also gunner, and one driver. For these vehicles, including relief crews, 45 crews in all will be formed."

Nothing was said about the numbers of armored scout cars and tanks. As for the vehicles for which the crews were to be trained, there may have been some 25 newly built Steyr and circa 20 Hotchkiss and 25 Renaults, which were introduced new at the end of 1941 and beginning of 1942.

During 1942 some 30 Panhard armored scout cars were also taken over from the Wehrmacht.

On 8/18 the Reichsführer-Chief sent a message to the Chief of the Ordnungspolizei:

"Through the intervention of the Reich Security Headquarters, SS Group Leader Streckenbach, we have received to date 13 armored scout cars from the Wehrmacht for the purpose of fighting bandits. I

order that of these armored scout cars 3 go to Minsk, 5 to SS Ogrf von dem Bach, 5 to SS Ogrf Prützmann. They will be directed and apportioned to the Ordnungspolizei. SS Ogrf. von dem Bach and SS Ogrf. Prützmann will each receive from me personally the assignment to turn over at least two of these armored scout cars in the [p. 216] next months to the Sipo, to gain practical experience in their informational activities of locating the bandits. About the use of the armored scout cars in service you are to report to me by 12/31/42."

If these instructions were followed, then three armored scout cars were apportioned to the HSSPF. North (Jeckeln) for Minsk (White Ruthenia), five to the HSSPF. Center (v.d.Bach) and five to the HSSPF. South (Prützmann). Unfortunately it is not clear what type of armored scout cars they were.

On May 29, 1943 the deputy of the Reichsführer wrote, under the deputy of the Führer, General of the Infantry von Unruh, to the Ordnungspolizei headquarters in Berlin, saying that there was progress made in obtaining two or three armored cars from Army High Command In 6 for fighting bandits in Oberkrain. The Inspector General of the Armored Troops had reported be teletype that Army High Command/Chief H Rüst/ AHA/In 6 was advised to assign ten Hotchkiss tanks. These should be taken from the armored supply depot in Gien on the Loire, some 160 km southeast of Paris. At the time it was not clear whether enough short barrels for the tank guns were on hand, since the majority of the tanks were probably fitted with long barrels, for which only very small quantities of ammunition were available. The change to short barrels should thus be made when possible.

According to this, ten Hotchkiss tanks could have reached the Ordnungspolizei in mid-1943.

During the Russian campaign captured Russian armored scout cars and tanks also came into Ordnungspolizei hands. Whether these armored vehicles were captured or salvaged by the Ordnungspolizei itself in its areas of action or assigned from the Wehrmacht is not clear.

After the Italian troops surrendered in September 1943, the Wehrmacht obtained great quantities of Italian armored vehicles. In addition, the production facilities of the Italian tank industry fell into German hands. Some of the captured vehicles and those produced from September 1943 on were assigned to the Ordnungspolizei. Here too, no complete overview is at hand, but there are several references to assignments and supplying.

As early as November 1943, in a report of the Inspector General of the Armored Troops on units of the Army Group Southeast that were equipped with captured Italian armored vehicles, whose equipping was ordered or was to take place shortly, four SS police regiments and their equipping with a total of twenty L6 tanks were mentioned.

The Supreme SSPF in Italy reported on February 18, 1944 that the BdO. Trieste was assigned four L 35/ 38 (Carette) tanks.

Further assignments to the Ordnungspolizei are found in reports of the Establishment Staff South of the Wehrmacht and the Command Staff of the HSSPF Italy. The Command Staff of the HSSPF Italy noted in their diary on September 1, 1944 that in the next two weeks or in the course of September it expected to receive eight P40 tanks, eight AB 43 scout cars and eight Lince scout cars. It can be seen in the monthly report of the Establishment Staff South for December 1944 that the Italian P40 tanks were available to the Supreme SS and Police Leader of Italy. From tank production in December 1944, all four finished P40 tanks were assigned to the hSSDPF Italy and two AB 43 armored scout cars to the BdO Croatia. For January 1944, production of ten P40 was expected, but by February 15, 1945 this had not been possible. New production of three P40 tanks by the end of March 1945 was planned.

Except for two P40 tanks that were given to the Army Weapons Office for testing, and five tanks that were used for training by the Armored Replacement and Training Unit South, the rest of the 61 vehicles produced by the end of 1944 could have been used by the Ordnungspolizei. At least Police Armored Companies 10 and 15 were equipped with 14 or 15 P40 tanks each before the war ended. In addition, in April 1944 half an armored company was listed by SSuPol.Fhr. Adria West; it too could have been equipped with P40 tanks.

Although no precise figures for the armored vehicles used by the Ordnungspolizei from 1936 to 1945 can be obtained, it becomes clear that there must have been well over 400 vehicles.

Model Descriptions

On the following pages, an overview of the various models of armored vehicles used by the Ordnungspolizei will be given, with selected technical data. The 46 pages of data will allow a minimal comparison of the individual models and give the numbers (often only an estimate) of them used by the Ordnungspolizei. As far as Ministry of the Interior numbers are known from documents or photos, they will likewise be given.

Krupp Armored Patrol Car

On the Krupp firm's chassis of Type L2H 143, the prototype of a modern armored police vehicle was created. The vehicle displayed as a Krupp Police Patrol Car at the 1938 Auto Show in Berlin was not used by the Police. The vehicle shows a certain similarity to the Wilton-Fijenoord armored car built in The Netherlands in 1933-34 on the Krupp L2H 43 chassis. (BWS)

EHRHARDT 1919 SPECIAL POLICE VEHICLE

Pskw. Ehrhardt 1919 with weapons removed, all machine-gun loopholes open, equipment such as jack, loader and crowbar carried on the vehicle.

Technical Data:

Weight (tons):	7.7	Armor (mm):	4-9
Length (cm):	530	Engine type:	Otto 4-cylinder
Width (cm):	200	Engine power (HP):	80
Height (cm):	290	Top speed (kph):	61
Crew (number):	7-8	Range (km):	250

Armament: One 7.92 mm MG in turret, one 7.92 mm MG usable elsewhere in body
Ammunition: 10,000 belted rounds for machine guns
Manufactured in: Germany   by: Ehrhardt Automobilwerke
Numbers made: 12 1917 type + 20 1919 type   Years: 1917-1919
Original designation: Strassenpanzerwagen Ehrhardt 1917/19
Wehrmacht designation: none

Notes:
On the basis of a contract given by the Supreme Army Command on October 22, 1914, in 1915 one armored road car each was delivered for testing purposes by the firms of Daimler, Büssing and Ehrhardt. The successful use of these vehicles led to a subsequent contract to Eberhardt for twelve vehicles, which were delivered with the designation of Strassenpanzerwagen Ehrhardt 1917 during World War I. After the war, the transitional army contracted for twenty more vehicles, which were introduced as Strassenpanzerwagen Ehrhardt 1919. All available road armored cars were taken over by various state police forces in 1920.
Number used by the Ordnungspolizei: 15
Confirmed Ministry numbers: see Appendix 1.

DAIMLER DZR SPECIAL POLICE VEHICLE

Daimler DZR Pskw. of the Baden State Police. The weapons are removed, the turret headlight flap opened.

Technical Data:

Weight (tons)	9	Armor (mm):	7-9
Length (cm):	590	Engine type:	Otto, 4-cylinder
Width (cm):	212	Engine power (HP):	100
Height (cm):	325	Top speed (kph):	45
Crew (number):	7-8	Range (km):	100-150

Armament: One 7/92 cm MG in turret, one 7/92 mm MG usable elsewhere in body
Ammunition: 10,000 belted rounds for machine guns
Manufactured in: Germany   By: Daimler
Number made: circa 40   Years: 1919-1920
Original designation: Strassenpanzerwagen Daimler DZR
Wehrmacht designation: unknown

Notes:

Along with the 20 Eberhardt armored cars, the transitional army also contracted with the Daimler firm for 40 armored road vehicles. These used the chassis of their reliable "Kr.-D 100 PS" towing tractor with an armored body and turning turret. The armored cars were intended for the Strassenpanzerwagen platoons of the temporary Reichswehr and transitional army, but because of the Treaty of Versailles, which forbade the Army to own armored vehicles, they had to be scrapped. Through negotiations with the Interallied Military Control Commission they received permission to use the armored cars in the state police forces. The Reichswehr kept a few vehicles illicitly.

Number used by the Ordnungspolizei: 25
Confirmed Ministry numbers: see Appendix 1.

BENZ V.P. TYPE 21 SPECIAL POLICE VEHICLE

Benz Pskw. 21 of the Bremen Police, with special signal-light apparatus for communication. (BWB)

Technical Data:

Weight (tons):	12	Armor (mm):	7-12
Length (cm):	595	Engine type:	Otto, 4-cylinder
Width (cm):	256	Engine power (HP):	100
Height (cm):	332	Top speed (kph):	48
Crew (number):	8-9	Range (km):	300-350
Armament:	two 7.92 mm MG 08 machine guns, each in its own turret		
Ammunition:	10,000 belted rounds for machine guns		
Made in:	Germany	by:	Benz & Cie.
Number made:	24	Years made: 1924-1925	
Original designation:	Sonderwagen für Polizeizwecke Benz		
Wehrmacht designation:	none		

Notes:

To equip state police forces with armored road cars, the Reich Ministry of the Interior coordinated the setting up of a commission with representatives of the states to work out the guidelines for the building of special police vehicles. This resulted in contracts to the Benz, Daimler and Ehrhardt firms. Following the guidelines resulted in almost identical vehicles that were equipped with two turning turrets for machine guns and an observation cupola for the commander. The chassis used had to have four-wheel drive plus forward and reverse steering. The Benz firm used the "VP 21" chassis.

Number used by the Ordnungspolizei: 24

Confirmed Ministry numbers: see Appendix 1

DAIMLER DZVR SPECIAL POLICE VEHICLE

The Daimler DZVR Pskw. without weapons. The flaps of the rear turret light are open.

Technical Data:

Weight (tons):	11	Armor (mm)	4-12
Length (cm):	610	Engine type:	Otto, 4-cylinder
Width (cm):	262	Engine power (HP):	100
Height (cm):	336	Top speed (kph):	50
Crew (number):	8-9	Range (km):	300

Armament: two 7.92 mm MG 08 machine guns, each in its own turning turret
Ammunition: 10,000 belted rounds for machine guns
Manufactured in: Germany by: Daimler
Number made: 33 Tears: 1924-1928
Original designation: Sonderwagen für Polizeizwecke Daimler
Wehrmacht designation: none

Notes:

To equip the state police forces with armored road vehicles, the Reich Ministry of the Interior as coordinating office set up a commission of state representatives, which worked out guidelines for the building of special police vehicles. This work resulted in contracts to the Benz, Daimler and Ehrhardt firms. Following the guidelines resulted in almost identically designed vehicles, which were fitted with two turning turrets for machine guns and an observation cupola for the commander. The chassis used had to have four-wheel drive plus forward and backward steering. The Daimler firm used the "DZVR" chassis.

Number used by the Ordnungspolizei: 26
Confirmed Ministry numbers: see Appendix 1

EHRHARDT 21 SPECIAL POLICE VEHICLE

Ehrhardt 21 Pskw. without weapons. Most of the equipment carried on the outside is also lacking.

Technical Data:

Weight (tons):	11	Armor (mm):	4-12
Length (cm):	650	Engine type:	Otto, 4-cylinder
Width (cm):	241	Engine power (HP):	80
Height (cm):	345	Top speed (kph):	56
Crew (number):	8-9	Range (km):	350

Armament: Two 7.92 mm MG 08 machine guns, each in its own turret
Ammunition: 10,000 belted rounds for machine guns
Manufactured in: Germany   by: Ehrhardt
Number made: 32   Years: 1923-1928
Original designation: Sonderwagen für Polizeizwecke Ehrhardt
Wehrmacht designation: none

Notes:

To equip the state police forces with armored road vehicles, the Reich Ministry of the Interior as coordinating office set up a commission with state representatives, which worked out the guidelines for building the special police vehicles. This work resulted in contracts to the Benz, Daimler and Ehrhardt firms. Following the guidelines resulted in vehicles almost identical in construction, fitted with two turning turrets for machine guns and an observation cupola for the commander. The chassis used had to have four-wheel drive plus forward and backward steering. The Ehrhardt firm used the "E-V" chassis.

Number used by the Ordnungspolizei: 32
Confirmed Ministry numbers: see Appendix 1

MAGIRUS SPECIAL POLICE VEHICLE

Magirus Pskw. of the Berlin Police. The turret with two machine guns is turned toward the back. (KM)

Technical Data:

Weight (tons):	8	Armor (mm):	secure against SmK
Length (cm):	520	Engine type:	unknown
Width (cm):	240	Engine power (HP):	unknown
Height (cm):	220	Top speed (kph):	60
Crew (number):	6	Range (km):	250
Armament:	Two 7.92 mm MG 08 machine guns in one turning turret		
Ammunition:	Unknown		
Manufactured in:	Germany	by:	Magirus
Number made:	1	Years:	1930-1932
Original designation:	Magirus Achtradfahrzeug		
Wehrmacht designation:	none		

Notes:

Since 1926 the state police forces, under the leadership of the Reich Ministry of the Interior, considered building a new medium special vehicle. The financial means for the police's own development were lacking, thus vehicles contracted for by the Reichswehr were adapted. In 1930, an eight-wheel vehicle was ordered from Magirus of Ulm, and in 1931 the chassis was tested in Baden in comparison with a Benz V.P. 21 special vehicle. The Magirus had advantages over the older type, but was found to be too expensive for series production. After completion in 1932, the vehicle was turned over to the Berlin Police.

Number used by the Ordnungspolizei: one

Confirmed Ministry number: 3981/32

BÜSSING Q31P SPECIAL POLICE VEHICLE

Büssing Q31P Pskw. of the Karlsruhe Police. The crew wear leather helmets. (ML)

Technical Data:

Weight (tons):	5.35	Armor (mm):	4-13
Length (cm):	557	Engine type:	Otto, 4-cylinder
Width (cm):	182	Engine power (HP):	60
Height (cm):	225	Top speed (kph):	70
Crew (number):	4	Range (km):	250-300

Armament: one 7.92 mm MG 08 machine gun in a turning turret
Ammunition: unknown
The technical data are based on the Büssing NAG G 31 P, Q 31 P chassis; there may be errors!

Manufactured in:	Germany	by:	Büssing NAG
Number made:	1	year:	1931

Original designation: Büssing NAG Sechsradwagen
Wehrmacht designation: Schwerer Panzerspähwagen (Sd.Kfz. 231)

Notes:

Since 1926 the state police forces, under the direction of the Reich Ministry of the Interior, considered building a new medium special vehicle. The financial means for developing it by the police were lacking, so the suitability of chassis and armored bodies developed for the Reichswehr was checked. In 1931 Baden, in cooperation with the Reich Ministry of the Interior, created a special vehicle on Büssing NAG chassis for testing by the police. The vehicle showed good driving qualities, but problems with turning the turret appeared from the start. The Q 31 P designation may be an error, since all Büssing chassis for this type of vehicle were G 31 P. Number used by the Ordnungspolizei: one
Confirmed Ministry number: none

DAIMLER L2000 SPECIAL POLICE VEHICLE

The Daimler L2000 Pskw. of the Karlsruhe Police was used as an armored personnel carrier.

Technical Data:

Weight (tons):	unknown	Armor (mm):	safe against SmK
Length (cm):	unknown	Engine type:	unknown
Width (cm):	unknown	Engine power (HP):	unknown
Height (cm):	unknown	Top speed (kph):	unknown
Crew (number):	unknown	Range (km):	unknown

Armament: One 7.92 mm MG 08 machine gun in turning turret
Ammunition: unknown

Manufactured in:	Germany	by:	Mercedes-Benz
Number made:	1	Year:	1931

Original designation: Sonderkraftwagen Mercedes-Benz 14/31
Wehrmacht designation: none

Notes:

Along with the Büssing Q 31 P, Baden obtained a Daimler special vehicle for testing in 1931. The test vehicles were types that had production chassis and were already contracted for by the Reichswehr. The vehicles were supposed to meet the minimum terms of tactical needs and be affordable in terms of available means. Their armament should consist of one or two machine guns in a 360-degree turning turret, the crew of two or three men. Scarcely any data on the vehicle delivered by Mercedes-Benz are available. The vehicle was used by the Karlsruhe Police for the protected transport of personnel and materials, but did not meet their minimum needs. Cracks in the armor appeared as early as 1932.

Number used by the Ordnungspolizei: one
Confirmed Ministry number: none

BÜSSING ARMORED CAR

This Büssing Pskw. was assigned to the Railroad Armored Platoon of the Army in 1939, but still bore the Pol-10019 number. (AWS)

Technical Data:

Weight (tons):	5.35	Armor (mm):	4-13
Length (cm):	557	Engine type:	Otto, 4-cylinder
Width (cm):	182	Engine power (HP):	60
Height (cm):	225	Top speed (kph):	70
Crew (number):	4	Range (km):	250-300
Armament:	One 2 cm tank gun and one 7.92 mm MG in turret		
Ammunition:	200 shells for 2 cm gun and 1500 rounds for machine gun		
Manufactured in:	Germany	by:	Büssing NAG
Number made:	1 (railcar)	Years:	1933-1935
Original designation:	Schwerer Panzerspähwagen (Sd.Kfz. 232)		
Wehrmacht designation:	Schwerer Panzerspähwagen (Sd.Kfz. 232)		

Notes:

The heavy six-wheel armored scout cars were developed in the early 1930s and the first vehicles delivered in 1933. Six-wheel armored scout cars were built by the firms of Daimler-Benz, Magirus and Büssing-NAG. One Büssing-NAG vehicle was equipped to run on rails, and was used as a railcar by Railroad Armored Platoon 3 of the Army in 1939. The armored scout car still bore the Ministry number Pol-10019 of the Munich Police Headquarters at that time. When the Munich Police obtained and later gave up this vehicle is not known. The 2 cm tank gun was not used in the car in 1939.

Number used by the Ordnungspolizei: one

Confirmed Ministry numbers: none (plate: Pol 10019)

STEYR ARMORED VEHICLE

A Steyr armored car of the 1935-37 series, in use by the 13th (strengthened) Police Armored Company, is seen on the Atlantic coast in 1943. (BAK)

Technical Data:

Weight (tons):	12	Armor (mm):	11
Length (cm):	626	Engine type:	Otto, 6-cylinder
Width (cm):	216	Engine power (HP):	150
Height (cm):	256	Top speed (kph):	70
Crew (number):	6	Range (km):	450
Armament:	One 2 cm tank gun and one 7.92 mm MG in turret, two 7.92 mm MG in body		
Ammunition:	100 shells for 2 cm gun and 5500 rounds for machine guns		
Manufactured in:	Austria	by:	Steyr-Daimler-Puch AG
Original designation:	Panzerwagen ADGZ		
Wehrmacht designation:	unknown		

Notes:

The development of this eight-wheel armored car was begun in Austria in 1931, and a prototype was built in 1933. Series production began in 1935, and the first 12 cars were delivered to the Bundesheer, which called this vehicle Panzerwagen M 42. Eight more were built for the Austrian Police and six for the Vienna Security Force. All the police vehicles were taken over by the Ordnungspolizei right after the "Anschluss", as were those of the Bundesheer later. The 25 cars produced in 1941-42 were equipped with MG 34 machine guns, with only one of them remaining in the body later.

Number used by the Ordnungspolizei: about 50

Confirmed Ministry numbers: 22770-22795 (1935-37), 20429-20444 (1941-42)

SKODA ARMORED CAR

Skoda "Turtle" armored car, in use by the Vienna Security Force. (MF)

Technical Data:

Weight (tons):	7.4	Armor (mm):	3-5.5
Length (cm):	600	Engine type:	Otto 4-cylinder
Width (cm):	216	Engine power (HP):	70
Height (cm):	244)	Top speed (kph):	70
Crew (number):	5	Range (km):	250
Armament:	four 7/62 mm machine guns in body		
Ammunition:	6250 rounds for machine guns		
Number made:	12	Years:	1924-1925
Original designation:	Panzerovy Automobil PA-II "Zelva" (OA vz. 23)		
Wehrmacht designation:	unknown		

Notes:

The Skoda PA II armored scout car developed out of a complete redesigning of the body of the predecessor PA I, of which two prototypes were built. The body of the PA II was especially striking because of its rounded shape and won the vehicle its nickname, "Turtle". The vehicles were never officially used by the Czechoslovakian armed forces, but used intensively in maneuvers and shown in parades. In 1927 three vehicles were sold to the Vienna Police; the other nine were turned over to the Czech police in 1937. One vehicle was displayed in Berlin on Wehrmacht Day.

Number used by the Ordnungspolizei: three (ex-Vienna Police)

Confirmed Ministry numbers: none

TATRA ARMORED CAR

Tatra Armored Car No. 1, without its weapons, during off-road testing. (AP)

Technical Data:

Weight (tons):	2.78	Armor (mm):	3-6
Length (cm):	402	Engine type:	Otto, 4-cylinder
Width (cm):	152	Engine power (HP):	32
Height (cm):	202	Top speed (kph):	60
Crew (number):	3	Range(km):	300

Armament: One 7.62 mm MG in turret, one MG in front body, one MG in reserve
Ammunition: 3000 rounds for machine guns
Manufactured in: Czechoslovakia   by: Tatra
Number made: 51   Years: 1933-1934
Original designation: OA vz. 30
Wehrmacht designation: Panzerspähwagen Tatra

Notes:

At the end of 1931 the Czechoslovakian Army decided to create the Tatra OA vz. 30 armored scout car, and the first vehicles were delivered in January 1934. The cars were used by various motorized or armored units. In February 1939 ten vehicles were turned over to the Czech police for support activities. After Czechoslovakia was dismembered in 1939, 18 vehicles remained with the Slovakian forces and nine in Romania. The Wehrmacht is believed to have captured 14 Tatra scout cars.

Number used by the Ordnungspolizei: 10 (ex-Czech police)
Confirmed Ministry numbers: 14035-14043

NETHERLANDS ARMORED CAR

An armored vehicle from the Netherlands in use by the 10th Heavy Company of Police Regiment Center, with name, number, national emblem and swastika. (BAK)

Technical Data:

Weight (tons):	6.1	Armor (mm):	9
Length (cm):	585	Engine type:	Otto, 6-cylinder
Width (cm):	210	Engine power (HP):	80
Height (cm):	250	Top speed (kph):	unknown
Crew (number):	5-6	Range (km):	unknown
Armament:	One 3.7 cm tank gun and one 7.92 mm MG in turret, two 7.92 mm MG in body		
Ammunition:	unknown		
Manufactured in:	Sweden	by:	Landsverk
Number made:	12	Years:	1935-1936
Original designation:	Landsverk L 181/Pantserwagen M 36		
Wehrmacht designation:	Panzerspähwagen L 202 (h)		

Notes:

In 1936 the Netherlands Army obtained 12 Swedish L 181 armored scout cars and introduced them as Pantserwagen M 36. Their chassis had been delivered to Landsverk from Mercedes-Benz. The 12 armored scout cars formed the 1st Eskadron Pantserwagen of the Netherlands Cavalry. It is not known how many M 36 were in condition to be used by the Wehrmacht, or how many of them were turned over to the Ordnungspolizei. Number used by the Ordnungspolizei: unknown, two vehicles documented
Confirmed Ministry numbers: none

KRUPP NETHERLANDS ARMORED CAR

The Krupp Netherlands Armored Car in service with the reinforced Armored Platoon Berlin, seen in the Chancellery yard in 1945. (TJ)

Technical Data:

Weight (tons):	4.5	Armor (mm):	unknown
Length (cm):	506	Engine type:	Otto, 4-cylinder
Width (cm):	220	Engine power (HP):	50
Height (cm):	230	Top speed (kph):	unknown
Crew (number):	3	Range (km):	unknown

Armament: one 7.92 mm MG in turret, two 7.92 mm MG in body, front and rear
Ammunition: unknown

Manufactured in:	Netherlands	by:	Wilton-Fijenoord N.V.
Number made:	3	Years:	1933-1934

Original designation: Wilton-Fijenoord Pantserwagen
Wehrmacht designation: unknown

Notes:

In 1933 and 1934 the Wilton-Fijenoord firm in the Netherlands built three armored cars on the three-axle Krupp L2H43 chassis. The firm had received the contract from the Royal Netherlands troops in the East Indies, but the vehicles proved to be too heavy for the road conditions in Java. Two vehicles were sold to Brazil, the third was in the Netherlands when the war began, was captured by the Wehrmacht and turned overt to the Ordnungspolizei.

Number used by the Ordnungspolizei: one
Confirmed Ministry numbers: none

PANHARD ARMORED CAR

Panhard with Ministry number 27164 in use by the 7th Police Armored Company in Russia in 1943. (KS)

Technical Data:

Weight (tons):	8.3	Armor (mm):	7-20
Length (cm):	514	Engine type:	Otto, 4-cylinder
Width (cm):	201	Engine power (HP):	105
Height (cm):	233	Top speed (kph):	72
Crew (number):	4	Range (km):	350

Armament: one 2.5 cm tank gun and one 7.5 mm machine gun in turret
Ammunition: 150 shells for tank gun, 3750 rounds for machine gun
Manufactured in: France   by: Panhard
Number made: about 480   Years: 1935-1940
Original designation: Automitrailleuse de Devouverte Panhard 178
Wehrmacht designation: Panzerspähwagen Panhard 204 (f)

Notes:

This very modern-looking armored scout car was adopted by the French Army as a standard vehicle in 1935. After the French campaign, the Wehrmacht made use of some 200 of these scout cars, and used a large number to supply the armored reconnaissance units of the 7th and 20th Armored Divisions. By the end of 1941, 109 Panhards were lost on the eastern front, and by mid-1942 the two reconnaissance units were disbanded or reequipped with German products. The remaining Panhards were used by reconnaissance and securing units, were rebuilt as railcars, or were turned over to the Ordnungspolizei.

Number used by Ordnungspolizei: about 30
Confirmed Ministry numbers: 27163 to 27187

RUSSIAN LIGHT ARMORED CAR

A light Russian BA-20 armored car in service with the Feldgendarmerie of the Wehrmacht in Russia in mid-1944. (BAK)

Technical Data:

Weight (tons):	2.5	Armor (mm):	9-10
Length (cm):	431	Engine type:	Otto, 4-cylinder
Width (cm):	175	Engine power (HP):	50
Height (cm):	213	Top speed (kph):	85
Crew (number):	3	Range (km):	450
Armament:	One 7.62 mm machine gun in turret		
Ammunition:	1386 rounds for machine gun		
Manufactured in:	Soviet Union	by:	Vykunskiy Werk
Number made:	unknown	Years:	1936-1939?
Original designation:	Light Armored Scout Car BA-20		
Wehrmacht designation:	Panzerspähwagen BA 202 (r)		

Notes:

Russia had gained experience with armored vehicles in the Spanish Civil War, where native armored cars, built on available truck chassis, and vehicles imported from Great Britain were used. Early in the 1930s a number of light armored scout cars on two-axle chassis were developed, and were replaced as of 1936 by the BA-20 type on GAZ-M1 chassis. There were a command version BA-20V with radio equipment, at first with a striking frame antenna, and a rail version BA-20ZhD. Captured BA-20 scout cars were also used by the Wehrmacht.

Number used by the Ordnungspolizei: unknown, two documented

Confirmed Ministry numbers: none

LIGHT RUSSIAN ARMORED CAR

A light Russian BA-64 armored car, such as was used by the 5th (reinforced) Police Armored Company.

Technical Data:

Weight (tons):	2.4	Armor (mm):	6-10
Length (cm):	366	Engine type:	Otto, 4-cylinder
Width (cm):	152	Engine power (HP):	50
Height (cm):	190	Top speed (kph):	80
Crew (number):	2	Range (km):	450

Armament: One 7.62 mm machine gun
Ammunition: 1070 rounds for the machine gun
Manufactured in: Soviet Union   by: GAZ
Number made: about 3500   Years: 1941-1945
Original designation: BA-64 or 64B armored scout car
Wehrmacht designation: Panzerspähwagen BA-64 (r)

Notes:

In 1941 the BA-64 armored scout car was developed on the chassis of the light GAZ-64 off-road vehicle (Russian Jeep); small numbers of them were built in the same year and 1942. In 1943 an improved version, the BA-64B, appeared on the chassis of the GAZ-64B off-road car; many more of these were produced. In the BA-64 the machine gun was mounted on a pivot in the open fighting compartment, while in the BA-64B it was housed in a small open-topped turret. Since the Wehrmacht was already drawing back in 1943, only a few of these light armored scout cars were captured and used by the Germans.

Number used by the Ordnungspolizei: unknown; one documented

Confirmed Ministry numbers: none

HEAVY RUSSIAN ARMORED CAR

A heavy Russian armored car used by the Ordnungspolizei, photographed in Poniatova, Poland, in early 1944. (JF)

Technical Data:

Weight (tons):	5.1	Armor (mm):	6-15
Length (cm):	465	Engine type:	Otto, 4-cylinder
Width (cm):	207	Engine power (HP):	50
Height (cm):	221	Top speed (kph):	55
Crew (number):	4	Range (km):	300

Armament: one 4.5 cm tank gun and one 7.62 mm MG in the turret, one 7.62 mm MG in the body

Ammunition: 40 (49) shells for 4.5 cm tank gun, 2000 rounds for machine guns

Manufactured in:	Soviet Union	by:	Izhorskiy Werk
Number made:	unknown	Years:	1938-1941

Original designation: BA-3, BA-6 or BA-10 armored scout car

Wehrmacht designation: Panzerspähwagen BA 203 (r)

Notes:

In 1932 the development of heavy armored scout cars on three-axle chassis was carried on to replace the BA-27 armored scout car produced from 1928 to 1931. Production began with a small series of BA-1 on Ford-Timken chassis, with a 12.7 mm heavy machine gun or a 3.7 cm tank gun as the main turret weapon. The BA-1 was followed by the BA-3 on GAZ-AAA chassis, and the BA-6 and BA-10 on GAZ-M1 chassis. The BA-3 and BA-6 used turrets with 4.5 cm guns, such as were also used in T-26 and BT tanks. For the BA-10 a new turret with a 4.5 cm tank gun was developed.

Number used by the Ordnungspolizei: unknown; six documented

Confirmed Ministry numbers: none

ITALIAN AB 41 ARMORED SCOUT CAR

An Italian AB 41armored car of the Bozen Police Regiment is being loaded onto a ferry in 1944. (RE)

Technical Data:

Weight (tons):	7.5	Armor (mm):	6-8.5
Length (cm):	520	Engine type:	Otto, 6-cylinder
Width (cm):	192	Engine power (HP):	88
Height (cm):	248	Top speed (kph):	78
Crew (number):	4	Range (km):	320

Armament: 2 cm tank gun and 8 mm MG in turret, one 8 mm MG in body, pointing backward

Ammunition: 456 shells for tank gun, 1992 rounds for machine guns

Manufactured in:	Italy	by:	SPA
Number made:	about 650	Years:	1941-1944

Original designation: Autoblinda AB 41

Wehrmacht designation: Panzerspähwagen AB 41 201 (I)

Notes:

This standard armored scout car of the Italian Army was developed out of the previous Autoblinda AB 40 in 1941 by installing a more powerful engine and adding to the armament. When the Italian Army was disarmed in September 1943, the Wehrmacht took over 87 AB 41 scout cars, and another 23 were built for the Wehrmacht in 1944. After that the Wehrmacht began to produce the successor AB 43, with a stronger motor and changed turret, of which some 80 vehicles were delivered.

Number used by the Ordnungspolizei: unknown, four documented

Confirmed Ministry numbers: none

ITALIAN LANCIA ARMORED CAR

One of the oldest armored cars in Ordnungspolizei service was this Italian Lancia of the SS Police Regiment in Bozen. (RE)

Technical Data:

Weight (tons):	4.3	Armor (mm):	6-9
Length (cm):	540	Engine type:	Otto, 4-cylinder
Width (cm):	182	Engine power (HP):	36
Height (cm):	248	Top speed (kph):	60
Crew (number):	6	Range (km):	300
Armament:	Two 8 mm MG in turret, one 8 mm MG in rear body		
Ammunition:	15,000 rounds for machine guns		
Manufactured in:	Italy	by:	Lancia
Number made:	about 110	Years:	1917-1919
Original designation:	Autoblinda Lancia IZM		
Wehrmacht designation:	not known, probably Panzerspähwagen Lanzia IZM (I)		

Notes:

In 1915 there were 20 Lancia IZ built for the Italian Army. The vehicles had two turrets, one above the other, and were armed with three 6.5 mm Maxim machine guns, one in the upper and two in the lower turret. The successor model Lancia IZM had only one turret with two 8 mm St. Etienne machine guns, while its body was almost unchanged. The Lancia IZM Autoblinda was the standard type used by the Italian Army between the wars, and were still in action in the Balkans until 1943. The armament sometimes consisted of water-cooled and sometimes air-cooled 8 mm machine guns.

Number used by the Ordnungspolizei: unknown, probably just one.

Confirmed Ministry numbers: none

LINCE ARMORED VEHICLE

This Lince armored car, a copy of the Daimler Scout Car, was used by the 14th (reinforced) Police Armored Company. (BP)

Technical Data:

Weight (tons):	3.5	Armor (mm):	6-30
Length (cm):	324	Engine type:	Otto, 6-cylinder
Width (cm):	177	Engine power (HP):	70
Height (cm):	160	Top speed (kph):	80
Crew (number):	2	Range (km):	400

Armament: one 8 mm machine gun in left front body
Ammunition: 2000 rounds for machine gun
Manufactured in: Italy   by: Lancia
Number made: about 130 (105 in 1944)   Years: 1944-1945
Original designation: Autoblinda Lince
Wehrmacht designation: Panzerspähwagen Lince 202 (I)

Notes:

A copy of the British Daimler Scout Car planned by the Italian Army, no Autoblinda Lince was yet finished by Lancia in September 1943. The Wehrmacht contracted for 300 vehicles, 129 of which were probably delivered by the war's end. The Wehrmacht used the Lince scout cars in armored scout platoons (9 cars per platoon), sometimes also in mixed platoons with AB 41/43 scout cars. Lince scout cars were also used by staffs as armored communication vehicles.

Number used by the Ordnungspolizei: unknown, eight documented

Confirmed Ministry numbers: none

POLISH 7TP TRACKED TANK

The first tanks used by the Ordnungspolizei were these three Polish 7TP, seen here with Police Regiment Center in 1941. (BAK)

Technical Data:

Weight (tons):	9.9	Armor (mm):	5-18
Length (cm):	488	Engine type:	Diesel, 6-cylinder
Width (cm):	243	Engine power (HP):	110
Height (cm):	230	Top speed (kph):	32
Crew (number):	3	Range (km):	150
Armament:	one 3.7 cm tank gun and one 7.92 mm MG in turret		
Ammunition:	80 shells for tank gun and 3960 rounds for machine gun		
Manufactured in:	Poland	by:	Panztwowy Zaklad Inzynierii (PzInz)
Number made:	about 95	Years:	1937-1939
Original designation:	czolg lekki 7TP jednowiezowy		
Wehrmacht designation:	Panzerkampfwagen 7TP (p)		

Notes:

The 7TP tank was developed in Poland from Vickers six-ton tanks bought from Britain in the early 1930s. The primary weapons used were 3.7 cm tank guns made by the Swedish firm of Bofors, which also helped to develop the turret and produced the first turrets in Sweden. As of 1938, the turrets were produced in Poland. Until the war began, some 135 7TP tanks were produced, of which some 40 were equipped with twin turrets and called 7TP dw. Captured 7TP tanks were also used by the Wehrmacht.

Number used by the Ordnungspolizei: unknown; three documented

Confirmed Ministry numbers: none

RENAULT R35 TRACKED TANK

Renault R 35 tanks in service with the Police Armored Unit in Krainburg in 1942. (ML)

Technical Data:

Weight (tons):	9.8	Armor (mm):	14-40
Length (cm):	402	Engine type:	Otto, 4-cylinder
Width (cm):	185	Engine power (HP):	82
Height (cm):	210	Top speed (kph):	19
Crew (number):	2	Range (km):	138
Armament:	3.7 cm SA 18 tank gun and coaxial 7.5 mm machine gun in turret		
Ammunition:	102 shells for 3.7 cm tank gun and 2250 rounds for machine guns		
Manufactured in:	France	by:	Renault
Number made:	about 1600	Years:	1935-1940
Original designation:	Char leger Renault R 35		
Wehrmacht designation:	Panzerkampfwagen 35 R 731 (f)		

Notes:

The first contract for 300 vehicles was given in 1935. Exports before the war went to Yugoslavia, Poland, Romania and Turkey. Some 840 vehicles were overhauled by the Wehrmacht after the French campaign, but only a few were used as battle tanks. Most were used as tracked towing tractors and tank destroyers. In 1940 the Wehrmacht turned over 109 of them to Italy and 40 to Bulgaria. The Wehrmacht and Police fitted the originally closed observation cupolas on the turrets with two-part visors to give the commanders better sighting possibilities.

Number used by the Ordnungspolizei: unknown, about 25

Confirmed Ministry numbers: 19641-19647, 20578- 20881-20891

HOTCHKISS H 39 TRACKED TANK

Firing training with a Hotchkiss tank in 1942 at the Ordnungspolizei armor school in Vienna. (JF)

Technical Data:

Weight (tons):	12	Armor (mm):	12-45
Length (cm):	422	Engine type:	Otto, 6-cylinder
Width (cm):	185	Engine power (HP):	120
Height (cm):	214	Top speed (kph):	36
Crew (number):	2	Range (km):	150

Armament: One 3.7 cm SA 18 tank gun and one 7.5 mm MG in turret
Ammunition: 102 shells for tank gun and 2250 rounds for machine gun Manufactured in: France by: Hotchkiss
Number made: about 680 Years: 1938-1940
Original designation: Char leger Hotchkiss H 35 et H 39
Wehrmacht designation: Panzerkampfwagen 38 H 735 (f)

Notes:

The Hotchkiss H 39 tank was a further development of the H 35 produced since 1935, of which some 400 were built by 1930. Of the approximately 1100 H 35/38 tanks built, the Wehrmacht was able to overhaul some 600 for further use after the French campaign. Most of these tanks designated 38 H were fitted with the more powerful motor of the H 39, and some 500 had the long SA 38 tank gun. The Hotchkiss tank was used by the Army in special captured tank units. The captured Hotchkiss could be recognized by the rear spur and the two-part visor in the commander's cupola.

Number used by the Ordnungspolizei: about 20
Confirmed Ministry numbers: 19622 to 19639

PANZERKAMPFWAGEN I TYPE A

A Panzer I (Panzerkampfwagen I, Type A) in action with the 8th Police Armored Company. (BAK)

Technical Data:

Weight (tons):	5.4	Armor (mm):	5-13
Length (cm):	402	Engine type:	Otto, 4-cylinder
Width (cm):	206	Engine power (HP):	60
Height (cm):	172	Top speed (kph):	37
Crew (number):	2	Range (km):	140

Armament: Two 7.92 mm machine guns in turret
Ammunition: 2250 rounds for machine guns
Manufactured in: Germany   By: Krupp, Henschel, Daimler-Benz, etc.
Number made: about 1175   Years: 1935-1936
Original designation: LaS (Vs Kfz 617)
Wehrmacht designation: Panzerkampfwagen I Ausf. A (Sd.Kfz. 101)

Notes:

The Panzerkampfwagen I Type A was the Wehrmacht's first tank produced in large numbers. With this small vehicle the armored troops gained experience for later tactics in tank warfare. Actually intended as a quickly-produced vehicle for training, the tanks soon went into combat, first in the Spanish Civil War, then in Poland, Denmark, France and Africa. As of the autumn of 1941, Type A was withdrawn from the armored regiments and only used in the hinterlands and for training.

Number used by the Ordnungspolizei: unknown; one documented

Confirmed Ministry numbers: none

PANZERKAMPFWAGEN I TYPE F

This Panzer I (VK 1801) of the 2nd Police Armored Company (Ministry no. 28130) was in Aberdeen MD, USA in 1946. (TJ)

Technical Data:

Weight (tons):	21	Armor (mm):	25-80
Length (cm):	438	Engine type:	Otto, 6-cylinder
Width (cm):	264	Engine power (HP):	150
Height (cm):	205	Top speed (kph):	25
Crew (number):	2	Range (km):	150
Armament:	Two 7.92 mm machine guns in turret		
Ammunition:	5100 rounds for machine guns		
Manufactured in:	Germany	By:	Krauss-Maffei
Number made:	30	Year:	1942
Original designation:	Panzerkampfwagen I n.A. verstärkt		
Wehrmacht designation:	Panzerkampfwagen I Ausf. F (VK 1801)		

Notes:

The VK 1801 was a further development of the Panzer I with emphasis on stronger armor, contracted for at the end of 1939. With revised running gear and body, the vehicle bore little resemblance to the original Panzer I. The armor, up to 80 mm thick and scarcely to be penetrated by contemporary antitank weapons, raised the small tank's weight to 21 tons. No longer suited for combat by 1942 because of their fully insufficient armament, these tanks were used mostly in securing units.

Number used by the Ordnungspolizei: unknown; ten documented

Confirmed Ministry numbers: 28121 to 28130

PANZERKAMPFWAGEN II TANK

This Panzer II (VK 1601) was used by the 13th (reinforced) Police Armored Company. (TJ)

Technical Data:

Weight (tons):	19	Armor (mm):	20-80
Length (cm):	unknown	Engine type:	Otto, 6-cylinder
Width (cm):	unknown	Engine power (HP):	150
Height (cm):	unknown	Top speed (kph):	31
Crew (number):	3	Range (km):	unknown
Armament:	One 2 cm tank gun and one 7.92 mm machine gun in turret		
Ammunition:	unknown		
Manufactured in:	Germany	By:	MAN
Number made:	30	Years:	1940-1942
Original designation:	Panzerkampfwagen II n.A. verstärkt		
Wehrmacht designation:	Panzerkampfwagen II, Ausf. J (VK 1601)		

Notes:

The Panzerkampfwagen II (VK 1601), like the Panzer I (VK 1801), was contracted for at the end of 1939, with the emphasis on heavy armor. The two vehicles were very similar, with alternating road wheels and heavily armored hulls. The first test chassis of VK 1601 was finished in June 1940; the 30 vehicles of the zero series were delivered only in 1941 and 1942. Only a few of them saw front service, most being used by securing and training units.

Number used by the Ordnungspolizei: unknown; six documented

Confirmed Ministry numbers: none

PANZERKAMPFWAGEN III TANK

This Panzer III (Type F) was used by the 12th (reinforced) Police Armored Company. (MG)

Technical Data: Panzer III Type F (Type G):

Weight (tons):	19.8 (20.3)	Armor (mm):	10-35
Length (cm):	538 (541)	Engine type:	Otto, 12-cylinder
Width (cm):	291 (295)	Engine power (HP):	265
Height (cm(:	244	Top speed (kph):	67
Crew (number):	5	Range (km):	165
Ammunition:	3.7 cm tank gun and three 7.92 mm MG, or 5 cm tank gun and two 7.92 mm MG		
Ammunition:	121 tank gun shells and 4500 MG rounds, or 99 shells and 2700 rounds		
Manufactured in:	Germany	By:	Daimler-Benz, Henschel, MAN
Number made:	435 (600)	Years:	1939-1941
Original designation:	5. (6.) Serie ZW		
Wehrmacht designation:	Panzerkampfwagen III, Ausf. F (G) (Sd.Kfz. 141)		

Notes:

The Panzerkampfwagen III, developed since 1934, was the Wehrmacht's first battle tank that really deserved this title, as its design and armament suited it for tank-versus-tank combat. Type E was the first Type F that was produced in large numbers and had the final type of running gear. Type F differed very little from Type E. In Type G for the first time, the 5 cm tank gun was installed, and some older models with 3.7 cm tank guns were rearmed, so that mixed forms of the versions existed.

Number used by the Ordnungspolizei: unknown; three documented

Confirmed Ministry numbers: none

PANZERKAMPFWAGEN IV TANK

The heaviest German production tank used by the Ordnungspolizei was the Panzer IV Type F. (BAK)

Technical Data:

weight (tons):	22.3	Armor (mm):	10-50
Length (cm):	592	Engine type:	Otto, V-12-cylinder
Width (cm):	288	Engine power (HP):	265
Height (cm):	268	Top speed (kph):	42
Crew (number):	5	Range (km):	210

Armament: One 7.5 cm L/24 tank gun and one 7.92 MG in turret, one 7.92 mm MG in body.

Ammunition: 80 shells for tank gun and 3150 rounds for machine gun

Manufactured in:	Germany	By:	Krupp, Vomag, Nibelungenwerk
Number made:	470	Tears:	1941-1942

Original designation: Panzerkampfwagen IV (7.5 cm) (Sd.Kfz. 161) Ausführung F

Wehrmacht designation: same as above

Notes:

The Type F was a further step in the production of Panzer IV, which began in 1937 and was first developed in 1935 under the code name of "Begleitwagen" (B.W.). Unlike earlier types, its modifications included heavier bow (50 mm) and side (30 mm) armor, use of the Fahrersehklappe 30 visor, installation of the Kugelblende 50 mantlet for the bow machine gun, and modified running gear. Type F was the past Panzer IV production type with the short L/24 7.5 cm tank gun. After 470 had been built, the long 7.5 cm L/43 tank gun was installed. Number used by the Ordnungspolizei: four documented

Confirmed Ministry numbers: 18139 to 18141

BT 5 RUSSIAN TANK

A shot-down and captured BT 5 tank, recognizable by its angular riveted bow armor. (KM)

Technical Data:

Weight (tons):	11.5	Armor (mm):	6-13
Length (cm):	558	Engine type:	Otto, 120cylinder
Width (cm):	223	Engine power (HP):	350
Height (cm):	225	Top speed (kph):	72
Crew (number):	3	Range (km):	200

Armament: One 4.5 cm tank gun and one 7.62 mm machine gun
Ammunition: 115 shells for tank gun and 2394 rounds for machine gun
Manufactured in: Soviet Union By: Kharkov (KhPZ No. 183) Locomotive Factory
Number made: about 700 (all BT types) Years: 1933-1935
Original designation: Fast Tank BT-5
Wehrmacht designation: Panzerkampfwagen BT 742 (r)

Notes:
The Russian BT tanks were developed from the American Christie M1930, two of which were imported in 1931 and designated BT-1. The first production series, which began in 1931, was called BT-2. This fast tank could run on tracks or its four large road wheels, and could be converted in 30 minutes. There were two versions of the BT-2, one with three 7.62 mm machine guns and one with a 3.7 cm tank gun and one 7.62 mm machine gun in the turret. For the successor model BT-5 a bigger new turret which could hold the 4.5 cm tank gun was created.
Number used by the Ordnungspolizei: unknown; one documented
Confirmed Ministry numbers: none

BT 7 RUSSIAN TANK

This BT 7 was used at Vienna-Purkersdorf in 1943 to train crews of Russian tanks. (FK)

Technical Data:

Weight (tons):	14	Armor (mm):	6-13
Length (cm):	566	Engine type:	Otto, 12-cylinder
Width (cm):	229	Engine power (HP):	500
Height (cm):	242	Top speed (kph):	86
Crew (number):	3	Range (km):	250

Armament: One 4.5 cm tank gun and one 7.62 mm machine gun in turret
Ammunition: 146 shells for tank gun and 2394 rounds for machine gun
Manufactured in: Soviet Union   By: Kharkov (KhPZ No. 183) Locomotive Factory
Number made: about 7000 (all BT types)   Years: 1935-1939
Original designation: Fast Tank BT-7
Wehrmacht designation: Panzerkampfwagen BT 742 (r)

Notes:

After changes to the body and installation of a new engine (BMW copy), the Type BT-7 emerged in 1935. In 1937 a new turret was introduced, offering better ballistic protection with its angled armor. Otherwise the vehicle was almost identical and was still designated BT-7. Some later-series turrets had another machine gun in the back. Further improvements (such as a Diesel engine) led to the BT-8 in 1939, production of which ended in 1941. The Wehrmacht formed several captured tank units armed with BT-5 and BT-7 tanks in 1941-42.

Number used by the Ordnungspolizei: unknown; two documented

Confirmed Ministry numbers: none

T 26 RUSSIAN TANK

Rear view of a T-26 (Model 33) than ran into a twin-turret T-26 (Model 31). (BWS)

Technical Data: T-26 Model 33 (T-26S Model 37):

Weight (tons):	9.4 (10.5)	Armor (mm):		6-25
Length (cm):	488	Engine type:		Otto, 4-cylinder
Width (cm):	341	Engine power (HP):		91
Height (cm):	241	Top speed (kph):		28 (30)
Crew (number):	3	Range (km):		175 (225)
Armament:	One 4.5 cm tank gun and one 7.62 mm machine gun in turret			
Ammunition:	100 (165) shells for tank gun and 3000 rounds for machine gun			
Manufactured in:	Soviet Union	By:		Zavod No. 174, 185 and others
Number made:	about 12,000 (att T-26 types)		Years: 1931-1941	
Original designation:	Panzerkampfwagen T-26			

Wehrmacht designation: Panzerkampfwagen T-26A 737 (r), T-26B 738 (r), T-26C 740 (r)

Notes:

The T-26 tank was developed from imported Vickers-Armstrong tanks in 1930. Production began in 1931 with 120 double-turret T-26 Model 31 types, with either one machine gun per turret or one machine-gun turret and one for a 4,5 cm tank gun and coaxial 7.62 mm machine gun. At the end of Model 33 production, turrets with two more machine guns, in the rear and for anti-aircraft use, were used. There followed the T-26S Model 37 (new turret) and, in 1939, the T-26S Model 39 with a new turret and body.

Number used by Ordnungspolizei: unknown; seven documented

Confirmed Ministry numbers: none

T 60 RUSSIAN TANK

The light T 60 tank of an unknown Ordnungspolizei unit in Russia. (KM)

Technical Data: Model 41 (Model 42):

Weight (tons):	5.8 (6.4)	Armor (mm):	7-20 (7-35)
Length (cm):	410	Engine type:	Otto, 6-cylinder
Width (cm):	230	Engine power (HP):	70 (85)
Height (cm):	174	Top speed (kph):	44 (45)
Crew (number):	2	Range (km):	450

Armament: One 2 cm rapid-fire tank gun and one 7.62 mm MG in turret
Ammunition: 750 (780) shells for tank gun plus machine-gun ammunition
Manufactured in: Soviet Union By: Zavol No. 37, 38, and GAZ
Number made: about 6000 Years: 1941-1942
Original designation: T-60 Light Tank
Wehrmacht designation: Panzerkampfwagen T-60 743 (r)

Notes:

The production of the T-60 light tank began in July 1941. As of about mid-1942 several changes were made to it. The road wheels and return rollers, originally spoked, were now made as disc wheels. A somewhat more powerful engine of the GAZ 203 type was installed, and armor plate was added to the hull and turret, bringing the thickness up to 35 mm on the bow and 25 mm on the sides. In September 1942 production was halted, as the T-60 had not proved itself. The vehicles were too slow, too weakly armored, and insufficiently armed for tank combat.

Number used by the Ordnungspolizei: unknown; seven documented

Confirmed Ministry numbers: none

T 70 RUSSIAN TANK

These two T 70 tanks saw service with Workshop Platoon 14 of the 5th (reinforced) Police Armored Company. (JF)

Technical Data:

Weight (tons):	9.2	Armor (mm):	10-60
Length (cm):	429	Engine type:	two Otto 6-cylinder
Width (cm):	232	Engine power (HP):	2 x 70
Height (cm):	204	Top speed (kph):	45
Crew (number):	2	Range (km):	360

Armament: One 4.5 cm tank gun and one 7.62 mm machine gun in turret
Ammunition: 94 shells for tank gun and 845 rounds for machine gun

Manufactured in:	Soviet Russia	By:	Zavon No. 37, 38, and GAZ
Number made:	about 8200	Years:	1942-1943
Original designation:	T-70 Tank		
Wehrmacht designation:	Panzerkampfwagen T-70 (r)		

Notes:

From the efforts to improve the handling, weak armor and light armament of the T-60 there resulted the T-70 tank, production of which began in March 1942. The use of two engines, mounted one behind the other on the right side and each driving one set of running gear, was interesting. As in the T-60, the commander in the turret was overworked, operating the guns as well as commanding the tank. T-70 production ended in October 1943. Captured T-60 and T-70 tanks were often used by the Wehrmacht as tracked towing tractors with their turrets removed.

Number used by the Ordnungspolizei: unknown; three documented

Confirmed Ministry numbers: none

T 34 RUSSIAN TANK

This T 34 (Model 43) tank served with the 5th (reinforced) Police Armored Company in 1944. (JF)

Technical Data: Model 42 (Model 43):

Weight (tons):	28.5 (30.9)	Armor (mm):	20-65 (20-70)
Length (cm):	668 (675)	Engine type:	Diesel, V-12-cylinder
Width (cm):	300	Engine power (HP):	500
Height (cm):	245	Top speed (kph):	55
Crew (number):	4	Range (km):	400 (465)

Armament: One 7.62 cm tank gun and one 7.62 mm MG in turret, one 7.62 mm MG in bow

Ammunition: 77 (100) shells for tank gun and 2000 to 3000 rounds for MG

Manufactured in:	Soviet Union	By:	Zavod No. 112, STZ, UZTM, etc.
Number made:	about 35,000	Years:	1940-1944
Original designation:	T-34 Tank		
Wehrmacht designation:	Panzerkampfwagen T-34 747 (r)		

Notes:

Production of the T-34 began in 1940, and the outstanding combination of off-road capability, armor and firepower made it one of the most successful battle tanks of World War II. The models built in 1940 were still armed with the weaker 7.62 cm L-11 tank gun and had various technical problems. From 1941 on the 7.62 cm F-34 tank gun was used, remaining the primary weapon until 1944. The vehicles from various manufacturers differed slightly in the form of the turrets and road wheels. The most striking change was the use of a roomier turret in the 1943 model, sometimes called T-43.

Number used by the Ordnungspolizei: unknown; 12 documented

Confirmed Ministry numbers: none

VICKERS MARK VI TANK

A British Vickers Mark VIB tank captured in the western campaign. (KM)

Technical Data:

Weight (tons):	5.3	Armor (mm):	8-15
Length (cm):	389	Engine type:	Otto, 6-cylinder
Width (cm):	205	Engine power (HP):	88
Height (cm):	223	Top speed (kph):	56
Crew (number):	3	Range (km):	208
Armament:	One 12.7 mm and one 7.7 mm machine gun in turret		
Ammunition:	unknown		
Manufactured in:	Great Britain	By:	Vickers
Number made:	unknown	Years:	1936-1940
Original designation:	Light Tank Mark VI		
Wehrmacht designation:	le. Panzerkampfwagen Mk VIB 735 (e)		

Notes:

The Vickers Mark VI light tank was the last developmental stage of a series that had begun in 1929 with the Mark I. The vehicle was developed from tanks made by Carden-Lloyd, which was taken over by Vickers in 1928. The Vickers Mark VI light tank was in active service with the British troops until 1941 and was also exported to other countries. The Wehrmacht captured Vickers Mark VI tanks from British troops in the western campaign and in Africa.

Number used by the Ordnungspolizei: unknown; one documented

Confirmed Ministry numbers: 18246

VALENTINE MARK III TANK

This Mark III 749 (e) Infantry Tank was a test vehicle for the Army Weapons Office. (BAK)

Technical Data:

Weight (tons):	16.3	Armor (mm):	8-65
Length (cm):	541	Engine type:	Diesel, 6-cylinder
Width (cm):	263	Engine power (HP):	138
Height (cm):	227	Top speed (kph):	24
Crew (number):	3	Range (km):	144

Armament: One 4 cm tank gun and one 7.92 mm machine gun in turret
Ammunition: 79 shells for tank gun and 3150 rounds for machine gun
Manufactured in: Britain, Canada   By: Vickers-Armstrong, Canadian Pacific
Number made: 8275 (all types)   Years: 1939-1944
Original designation: Tank, Infantry, Mk. III Valentine
Wehrmacht designation: Infanterie Panzerkampfwagen Mk III 749 (e)

Notes:

This tank owes its name to Valentine's Day, since it was first displayed to the British Ministry of War on February 14, 1938 (Valentine's Day). This vehicle was a "private" development of the Vickers firm. 2394 British and 1388 Canadian Valentines were delivered to the Soviet Union in the Lend-Lease agreement, some 30% of British and up to 30 tanks of the entire Canadian production. Because of its technical reliability, the Valentine was very popular among its Russian crews; only the 4 cm tank gun was too weak.

Number used by the Ordnungspolizei: unknown; two documented

Confirmed Ministry numbers: none

L 35 ITALIAN TANK

This L 35 tank was turned over to the 1st Company, SS Police Regiment Bozen in 1944. (BAK)

Technical Data:

Weight (tons):	3.4	Armor (mm):	6-13.5
Length (cm):	314	Engine type:	Otto, 4-cylinder
Width (cm):	140	Engine power (HP):	43
Height (cm):	128	Top speed (kph):	41
Crew (number):	2	Range (km):	120
Armament:	Two parallel 8 mm machine guns in body		
Ammunition:	2400 rounds for machine guns		
Manufactured in:	Italy	By:	Fiat-Ansaldo
Number made:	about 1600 (all types)	Years:	1933-1944
Original designation:	Carro Armato L.3-35		
Wehrmacht designation:	Panzerkampfwagen L/3-35 731 (i) [small i]		

Notes:

This small tank was developed from the British Carden-Lloyd Mark VI. The Type L 3.33, produced since 1933, differed only in minor details from the Tyle L 3.35 that followed it in 1935. The vehicles were exported to many countries, including 74 to Austria, where they were taken over by the Wehrmacht in 1938. A flame throwing version was also produced, and in 1942-43 84 vehicles of an improved type had modified chassis. In September 1944 the Wehrmacht took over 148 tanks, and 17 more were built that year. They were used in light tank platoons of the Infantry, for securing tasks and by the Organization Todt.

Number used by the Ordnungspolizei: unknown; 14 documented

Confirmed Ministry numbers: none

L 6 ITALIAN TANK

An L 6 tank, seen in service with riflemen of SS Police Mountain Jäger Regiment 18 in 1944.

Technical Data:

Weight (tons):	6.8	Armor (mm):	6-30
Length (cm):	383	Engine type:	Otto, 4-cylinder
Width (cm):	185	Engine power (HP):	70
Height (cm):	218	Top speed (kph):	42
Crew (number):	2	Range (km):	200
Armament:	One 2 cm tank gun and one 8 mm machine gun in turret		
Ammunition:	312 shells for tank gun and 1560 rounds for machine gun		
Manufactured in:	Italy	By:	Fiat
Number made:	about 450	Years:	1939-1944
Original designation:	Carro Armato L/6		
Wehrmacht designation:	Panzerkampfwagen L 6 733 (i)		

Notes:

This tank was developed from a Fiat-Ansaldo design intended for export and based on an improved L 3 chassis. In 1936 the first prototypes were finished, but production began only in 1939. Besides the standard version, there were also a command tank and a flame throwing version. Of the approximately 422 originally built, the Wehrmacht took over 106 in September 1943. In 1943-44 another 17 L 6 (including two command tanks) were built for the Wehrmacht. They were used mainly in reconnaissance platoons (of five tanks each) of infantry divisions.

Number used by the Ordnungspolizei: up to 20 assigned Nov. 1944, 5 documented

Confirmed Ministry numbers: none (registered Pol-228494)

M 15 ITALIAN TANK

Carro Armato M 15/42, used by the Ordnungspolizei as K-Pzkw. M 15. (BAK)

Technical Data:

Weight (tons):	15	Armor (mm):	14.5-30
Length (cm):	504	Engine type:	Diesel, V-8-cylinder
Width (cm):	223	Engine power (HP):	185
Height (cm):	239	Top speed (kph):	38
Crew (number):	4	Range (km):	180

Armament: One 4.7 cm tank gun and one 8 mm MG in turret, two 8 mm MG in bow, one 8 mm AA-MG

Ammunition: 87 shells for tank gun, 2736 rounds for machine guns

Manufactured in:	Italy	By:	Fiat-Ansaldo
Number made:	220	Years:	1942-1943

Original designation: Carro Armato M 15/42

Wehrmacht designation: Panzerkampfwagen M 15/42 738 (i)

Notes:

This was the result of developing a tank that was begun in 1937 and led via the M 11/39, M 13/40 and M 14/41 to the M 15/42. Unlike the M 14/41, the M 15/42 had a more powerful motor, an improved tank gun, and electric turret turning apparatus. Changes to the body were minimal. In September 1943 the Wehrmacht took over 147 M 13/40, 14/41 and 15/42, which were used to equip whole captured tank companies and units, including some in the Waffen-SS. All these units served in the Balkan area. In 1944, 28 more M 15/42 were built.

Number used by the Ordnungspolizei: unknown; five documented

Confirmed Ministry numbers: none

P 40 ITALIAN TANK

The P 40 tank prototype shown here differed in several details from the production model (see pp. 125, 147 and 157). (BAK)

Technical Data:

Weight (tons):	26	Armor (mm):	14-50
Length (cm):	580	Engine type:	Diesel, V-8-cylinder
Width (cm):	280	Engine power (HP):	330
Height (cm):	250	Top speed (kph):	42
Crew (number):	4	Range (km):	275

Armament: One 7.5 cm L/34 tank gun and one 8 mm MG in turret, one 8 mm AA-MG
Ammunition: 65 shells for tank gun and 576 rounds for machine guns

Manufactured in:	Italy	By:	Ansaldo
Number made:	61 with motors	Years:	1944-45

Original designation: Carro Armato P 40
Wehrmacht designation: Panzerkampfwagen P 40 737 (i)

Notes:

The Italian Army had already contracted in April 1942 for 500 of these tanks, which were based strongly on the T 34. But in 1942 only one prototype was built, and nothing else was done until September 1943. The Wehrmacht then contracted for 150 tanks, of which 101 were delivered in 1944. Because of problems with the engine, the Diesel was to be replaced by a V-12 gasoline engine, but only 61 tanks were delivered with engines, the rest being used without motors as fixed gun emplacements. Wehrmacht plans to use the tanks as artillery vehicles were given up for poor engine and weapon performance.

Number used by the Ordnungspolizei: 55 at most; 29 documented

Confirmed Ministry numbers: none

L 6 ITALIAN ASSAULT GUN

Several Semovente da 47/42 were used by the Ordnungspolizei as L 6 assault guns. (BAK)

Technical Data:

Weight (tons):	6.5	Armor (mm):	6-30
Length (cm):	383	Engine type:	Otto, 4-cylinder
Width (cm):	185	Engine power (HP):	70
Height (cm):	172	Top speed (kph):	42
Crew (number):	3	Range (km):	200

Armament: One 4.7 cm gun in body, sometimes one 8 mm MG behind shield
Ammunition: 49 shells for 4.7 cm gun plus machine-gun ammunition

Manufactured in:	Italy	By:	Fiat
Number made:	about 300	Years:	1941-44

Original designation: Semovente da 47/32 (Scafo L 40)
Wehrmacht designation: Sturmgeschütz L 6

Notes:

This self-propelled mount for the 4.7 cm antitank gun on the L6 tank chassis was introduced in 1941. The vehicle was never intended as a tank destroyer by the Italian Army, but as a light attack and support vehicle (assault gun). It was supposed to operate along with light tank or reconnaissance tanks or against fortified positions. The 35 taken over by the Wehrmacht as of September 1943 and the 74 built new afterward also saw service in reconnaissance and light tank units. Some were also used as towing tractors for antitank guns.
Number used by the Ordnungspolizei: unknown; 25 documented
Confirmed Ministry numbers: none (registration: Pol-228.)

M 42 ITALIAN ASSAULT GUN

Captured Semovente M 42 da 75/18 were used as Assault Gun M 42. (PGS)

Technical Data:

Weight (tons):	14	Armor (mm):	14.5-50
Length (cm):	507	Engine type:	Otto, V-8-cylinder
Width (cm):	245	Engine power (HP):	185
Height (cm):	174	Top speed (kph):	38
Crew (number):	3	Range (km):	180
Armament:	One 7.5 cm L 18 gun and one 8 mm AA machine gun		
Ammunition:	44 shells for gun and 1334 rounds for machine gun		
Manufactured in:	Italy	By:	Ansalso
Number made:	245	Years:	1942-44
Original designation:	Semovente M 42 da 75/18		
Wehrmacht designation:	Sturmgeschütz M 42 mit 75/18 851 (i)		

Notes:

In 1941, 60 assault guns with 7.5 cm L 18 guns were built on the chassis of the M 13/40 tank. In 1942 another 162 were built on the M 14/41 chassis. Of the 293 ordered on the M 15/42 chassis, two were delivered in 1942 and another 188 by September 1943. The Wehrmacht took over 131 in September 1943 and had another 55 built in 1944. These Italian assault guns were used to form assault-gun units and tank-destroyer companies in Italy and the Balkans.

Number used by the Ordnungspolizei: unknown; nine documented

Confirmed Ministry numbers: none

FLAMETHROWING TANKS

A flame throwing armored car of the 13th (reinforced) Police Armored company, seen on the Atlantic coast in mid-1944.

Technical Data:

Weight (tons):	8.62	Armor (mm):	5.5-14.5
Length (cm):	580	Engine type:	Otto, 6-cylinder
Width (cm):	210	Engine power (HP):	100
Height (cm):	210	Top speed (kph):	50
Crew (number):	4	Range (km):	300
Armament:	Two 14 mm jets, one 7 mm mobile jet, one 7.62 mm machine gun		
Ammunition:	700 liters of burning oil for jets, 2010 rounds for machine gun		
Manufactured in:	Germany	By:	Hanomag/Wumag
Number made:	over 300?	Years:	1943-44
Original designation:	mittlerer Flammpanzer (Sd.Kfz. 251/16)		
Wehrmacht designation:	same as above		

Notes:

The production of the medium flame throwing armored car (Sd.Kfz. 251/16) began in January 1943. On the medium Schützenpanzerwagen (Sd.Kfz. 251), a flame throwing apparatus made by the Koebe firm, with two 14 mm jets behind armor shields, and one 7 mm jet on a ten-meter hose (with additional ten-meter lengthening) were built. The 14 mm jets had a traversing radius of 160 degrees; the 7 mm jet was carried on the rear body in a special bracket. The 700 liters of oil were enough for 80 blasts (one second each) with a range of 50 to 60 meters. The Wehrmacht formed flame platoons (gp) of six vehicles each.

Number used by the Ordnungspolizei: two documented

Confirmed Ministry numbers: 28142 and 28143

TL 37 ITALIAN PERSONNEL CARRIER [alias Panzer 10]

A TL 37 250 (i) armored personnel carrier used by the 7th SS Volunteer Mountain Division "Prinz Eugen". These vehicles were also used by the Ordnungspolizei as Panzer 10 (ital) because they could transport ten people. (BAK)

Technical Data:

Weight (tons):	5.3	Armor (mm):	6-8.5
Length (cm):	495	Engine type:	Otto, 4-cylinder
Width (cm):	192	Engine power (HP):	67
Height (cm):	213	Top speed (kph):	52
Crew (number):	8-10	Range (km):	600
Armament:	One 8 mm machine gun		
Ammunition:	unknown		
Manufactured in:	Italy	By:	Ansaldo
Number made:	150	Tear:	1942
Original designation:	TL 37 Protetto		
Wehrmacht designation:	gepanzerter Mannschaftstransportwagen TL 37 250 (i)		

Notes:

In 1942 Ansalso produced 150 of these armored personnel carriers on the chassis of the light Fiat A.S. 37 "Desert Truck". Some of them were used by the Italian troops in the Balkans. When the Italian Army was disarmed in September 1943, the Wehrmacht took over 36 of the S 37 armored personnel carriers; no further production took place. In September 1943 the Ordnungspolizei began to use at least two of these vehicles, known as "Panzer 10 (ital)".

Number used by the Ordnungspolizei: unknown; two documented

Confirmed Ministry numbers: none

LIGHT ARMORED RAILCAR

Railcars of an armored train (le.Sp.) of the Wehrmacht. Such vehicles were also used by the Ordnungspolizei. (AWS)

Technical Data:

Weight (tons):	7.5	Armor (mm):	5.5-14.5
Length (cm):	569	Engine type:	Otto, V-8-cylinder
Width (cm):	unknown	Engine power (HP):	70
Height (cm):	unknown	Top speed (kph):	60
Crew (number):	5-6	Range (km):	400
Armament:	Four 7.62 mm machine guns		
Ammunition:	unknown		
Manufactured in:	Germany	By:	Steyr
Number made:	about 50	Years:	1943-45
Original designation:	leichter Schienenpanzer (spähwagen)		
Wehrmacht designation:	Panzerzug (le. Sp.)		

Notes:

By the end of the 1920s, Tatra had developed an armored railcar that could be used as an armored train, as a motorized single vehicle or with several cars coupled together. This idea was taken up by the German General Staff, which needed armored trains for use on rail lines on light foundations (such as in Greece and the Balkans). In August 1943 Steyr was contracted to build light armored railcars that could be coupled together into a train, and they were delivered beginning in the first half of 1944. The Wehrmacht always used ten of these railcars to form an armored train (le.Sp.), In action the vehicles traveled alone, or usually as pairs for mutual protection.

Number used by the Ordnungspolizei: unknown; four documented

Confirmed Ministry numbers: none

Along with photojournalists, there were also artists with Ordnungspolizei who were to portray the Ordnungspolizei's actions in newspapers for propaganda purposes. Here are two such drawings from the periodical "Die Deutsche Polizei": A Renault tank being refueled, April 7, 1943 (above), and a Tatra armored car in action in a Russian village in 1941 (below).

APPENDICES

Appendix 1

Distribution of Special Police Vehicles of the Ordnungspolizei in 1939

Maker	Type	Chassis #	Motor #	Model #	Base 1935-36 (1)	Repaired (2) 1/20/38	Sent to Berlin (3) 2/26/38	Also usable (4) 5/24/38	Used otherwise (5) 7/15/38	scrapped (6) 1/30/39
Benz	21	-	-	3050	Berlin					(+)
Benz	21	14016	14016	3051	Wilhelmsh.					x
Benz	21	14017	14017	6545	Hamburg					x
Benz	21	14018	14018	V	Stuttgart	x	x		Stuttgart	
Benz	21	14019	14019	VI	Stuttgart	x	x		Stuttgart	
Benz	21	14021	14021	IV	Stuttgart	x	x		Stuttgart	
Benz	21	14022	14022	2991	Dortmund					x
Benz	21	14023	14023	6814	München					x
Benz	21	14025	14025	2992	Essen					x
Benz	21	14026	14026	3005	Bochum					x
Benz	21	14027	14027	6813	München					x
Benz	21	14028	14028	3011	Düsseldorf					x
Benz	21	14030	14030	2999	Recklingh.	x	x		Berlin	
Benz	21	14031	14031	6812	München					x
Benz	21	14032	14032	3006	Wuppertal					x
Benz	21	14033	14033	3012	Gb.-Rheydt					x
Benz	21	14034	14034	-	Freiburg					x
Benz	21	14035	14035	-	Mannheim					x
Benz	21	16012	39418	6544	Hamburg					x
Benz	21	500/1		2990	Wiesbaden	x	x		(Berlin)	
Benz	21	500/15		3052	Kiel	x	x		Hamburg	
Benz	21	500/17		3010	Bochum	x	x		Berlin	
Benz	21	500/4	12104	6599	Darmstadt					x
Benz	21	500/9		3004	Frankfurt/M	x	x		Berlin	
Daimler	DZR	-	39318	2594	Breslau					x
Daimler	DZR	19693	39345	6443	Hamburg					x
Daimler	DZR	19694	39315	6444	Hamburg					x
Daimler	DZR	19698	39356	6445	Hamburg					x
Daimler	DZR	19700	39211	6446	Hamburg					x
Daimler	DZR	19701	39285	6447	Hamburg					x
Daimler	DZR	19715	39371	-	Chemnitz					x
Daimler	DZR	19716	39369	6513	Nürnberg					x
Daimler	DZR	19717	39303	-	Chemnitz					x

Maker	Type	Chassis #	Motor #	Model #	Base 1935-36 (1)	Repaired (2) 1/20/38	Sent to Berlin (3) 2/26/38	Also usable (4) 5/24/38	Used otherwise (5) 7/15/38	scrapped (6) 1/30/39
Daimler	DZR	19719	39253	-	Dresden					x
Daimler	DZR	19720	39360	-	Plauen					x
Daimler	DZR	19734	38928	6448	Hamburg					x
Daimler	DZR	19735	39393	-	Stuttgart					x
Daimler	DZR	19738	39338	-	Zwickau					x
Daimler	DZR	19740	39117	6514	Nürnberg					x
Daimler	DZR	19742	39202	6515	Nürnberg					x
Daimler	DZR	19751	39363	6759	Rostock					x
Daimler	DZR	19753	39133	-	Stuttgart					x
Daimler	DZR	19754	39240	2013	Berlin					x
Daimler	DZR	19756	39395	-	Leipzig					x
Daimler	DZR	19758	39410	6516	Nürnberg					x
Daimler	DZR	19759	39432	6517	Nürnberg					x
Daimler	DZR	19760	39412	-	Leipzig					x
Daimler	DZR	19761	39443	-	Stuttgart					x
Daimler	DZR	D.Z.1	39378	6600	Darmstadt					x
Daimler	DZVR	-	-	3000	Oppeln					x
Daimler	DZVR	13201	39501	3007	Essen					x
Daimler	DZVR	13202	-	2993	Berlin	x	x		Berlin	
Daimler	DZVR	13203	39493	3008	Stettin					x
Daimler	DZVR	13204	39505	2994	Hamburg					x
Daimler	DZVR	13205	-	2902	Breslau	x	x		Breslau	
Daimler	DZVR	13208	39524	2914	Elbing					x
Daimler	DZVR	13210	39492	6518	Nürnberg					x
Daimler	DZVR	13211	39491	6519	Nürnberg					x
Daimler	DZVR	13212	39420	3020	Wuppertal					x
Daimler	DZVR	13213	39585	3009	Schneidem					x
Daimler	DZVR	13214	39513	6520	Nürnberg					x
Daimler	DZVR	13215	-	2903	Hannover	x	x		Hannover	
Daimler	DZVR	13217	-	2916	Berlin	x	x		Berlin	
Daimler	DZVR	13219	-	2997	Potsdam	x	x		(?)	
Daimler	DZVR	13220	39449	6810	München					x
Daimler	DZVR	13222	39470	6070	Wilhelmsh.					x
Daimler	DZVR	13225	39516	6811	München					x
Daimler	DZVR	13227	-	2904	Gleiwitz	x	x		Breslau	
Daimler	DZVR	13228	-	3021	Waldenbg.	x	x		Breslau	

Maker	Type	Chassis #	Motor #	Model #	Base 1935-36 (1)	Repaired (2) 1/20/38	Sent to Berlin (3) 2/26/38	Also usable (4) 5/24/38	Used otherwise (5) 7/15/38	scrapped (6) 1/30/39
Daimler	DZVR	13230	39476	-	Braunschw.					x
Daimler	DZVR	14747	39463	3001	Hamburg					x
Daimler	DZVR	-	-	2871	Königsberg			x	Königsbg.	
Daimler	DZVR	-	-	2915	Königsberg			x	Königsbg.	
Daimler	DZVR	-	-	2995	Tilsit			x	Tilsit	
Daimler	DZVR	-	-	2996	Berlin					(+)
Ehrhardt	1919	-	-	1611	Weißenfels					x
Ehrhardt	1919	-	-	2339	Kiel					x
Ehrhardt	1919	912	947	3477	Duisburg					x
Ehrhardt	1919	916	951	-	Braunschw.					x
Ehrhardt	1919	917	952	-	Dessau					x
Ehrhardt	1919	918	935	1609	Berlin					x
Ehrhardt	1919	921	946	1612	Halle					x
Ehrhardt	1919	922	957	1610	Suhl					x
Ehrhardt	1919	925	960	2508	Gleiwitz					x
Ehrhardt	1919	926	1831	2507	Breslau					x
Ehrhardt	1919	927	962	-	Lübeck					x
Ehrhardt	1919	928	963	2227	Recklingh.					x
Ehrhardt	1919	930	? 930	3148	Berlin					x
Ehrhardt	1919	933	958	2789	Gleiwitz					x
Ehrhardt	1919	-	-	2039	Berlin					(+)
Ehrhardt	1921	2211	1199	3123	Berlin					x
Ehrhardt	1921	-	-	3055	Magdeburg					x
Ehrhardt	1921	2203	-	3045	Kassel	x	x		Hannover	
Ehrhardt	1921	2204	993	3002	Köln					x
Ehrhardt	1921	2205	-	3046	Stettin	x	x		Hamburg	
Ehrhardt	1921	2206	995	-	Chemnitz	x	x		(Berlin)	
Ehrhardt	1921	2207	926	6761	Rostock					x
Ehrhardt	1921	2208	-	2917	Dortmund	x	x		Köln	
Ehrhardt	1921	2209	-	2918	Duisburg	x	x		Essen	
Ehrhardt	1921	2210	1000	-	Weimar	x	x		(Berlin)	
Ehrhardt	1921	2211	-	3054	Hannover	x	x		Hannover	
Ehrhardt	1921	2212	-	2768	Erfurt	x		x	Erfurt	
Ehrhardt	1921	2213	-	3047	Oberhsn.	x		x	Oberhsn.	
Ehrhardt	1921	2214	-	2771	Düsseldorf	x	x		Essen	

Maker	Type	Chassis #	Motor #	Model #	Base 1935-36 (1)	Repaired (2) 1/20/38	Sent to Berlin (3) 2/26/38	Also usable (4) 5/24/38	Used otherwise (5) 7/15/38	scrapped (6) 1/30/39
Ehrhardt	1921	2315	1007	-	Dresden	x	x		Leipzig	
Ehrhardt	1921	2316	1006	-	Dresden	x	x		Leipzig	
Ehrhardt	1921	2317	1004	-	Leipzig	x	x		Leipzig	
Ehrhardt	1921	2318	-	3048	Recklingh.	x	x		Recklingh.	
Ehrhardt	1921	2219	-	2770	Frankfurt/M	x	x		Frankf./M.	
Ehrhardt	1921	3220	-	2919	Recklingh.	x	x		Recklingh.	
Ehrhardt	1921	2321	1010	6808	München					x
Ehrhardt	1921	2322	1008	-	Weimar	x	x		(Berlin)	
Ehrhardt	1921	2323	-	2772	Berlin	x	x		Berlin	
Ehrhardt	1921	2342	-	2788	Hamburg	x		x	Hamburg	
Ehrhardt	1921	2325	-	2769	Frankfurt/M	x	x		Frankf./M.	
Ehrhardt	1921	2326	1015	6809	München					x
Ehrhardt	1921	2327	-	3053	Magdeburg	x		x	Magdebg.	
Ehrhardt	1921	2328	-	3049	Aachen	x	x		Köln	
Ehrhardt	1921	2329	1018	6760	Rostock					x
Ehrhardt	1921	2330	1019	3124	Berlin					x
Ehrhardt	1921	3001	22/31	3003	Köln					x
Ehrhardt	1921	32482	-	3044	Halle/S.	x		x	Halle/S.	
Magirus	32	-	-	3981	Berlin					(+)
Büssing	Q31P	-	-	-	Karlsruhe	x	x		(Berlin)	
Merc.B.	14/33	-	-	-	Karlsruhe	x	x		(Berlin)	

(1) List compiled by author, using various sources, with no claim to completeness.
(2) According to Reichsführer-Chief's order (O-Kdo. T (2) 205 Nr. 2/38 of Jan. 30, 1938, of vehicles to be put into driveable condition.
(3) According to Reichsführer-Chief's order (O-Kdo. O (1) 6 Nr. 3/38 of Feb. 26, 1938) on "parade uses" of special armored vehicles sent to Berlin.
(4) According to Reichsführer-Chief's order (O-Kdo. T (2) 205 Nr. 21/38 of May 24, 1938) of special vehicles to be put into driveable condition but not sent to Berlin.
(5) According to Reichsführer-Chief's order (O-Kdo. T (2) 205 Nr. 23/38 of July 15, 1938) on special police vehicles to be sent to named police headquarters in driveable condition for training for armored car platoons.
(6) According to Reichsführer-Chief's order (O-Kdo. T (2) 205 Nr. 26/38 of Jan. 30, 1939) on special police vehicles to be scrapped.
(+) Returned to the Schutzpolizei by the state police forces in April 1935, fate unknown, maybe scrapped or turned over by unknown orders before January 1939.
(Berlin) Special vehicles on hand at the Police Headquarters in Berlin which remained on HQ lists.

Appendix 2

Strength Chart
Polizei M.G. Hundertschaft (mot.), 5/15/1938

3. (Polizei-Sonderkraftwagen) Zug
Kopfstärke: 1 Offizier, 3 Unterführer, 15 Männer, 8 Kraftfahrer
Waffen: 27 Pistolen, 4 Masch. Pistolen, 3 Karabiner, 4 schw. Maschinengewehre

Stkw. 4   Krad Bwg.   Krad   Krad   Krad   Mlkw. 23

Polizei-Sonderkraftwagen   Polizei-Sonderkraftwagen

Zum Teil waren die 3. Züge der Pol. M.G. Hdsch. mit drei PSkw. ausgestattet. Die Stärke erhöhte sich dann um 1 Unterf., 4 Männer 2 Kraftf., 7 Pistolen, 2 Masch. Pist., und 2 schw. Maschinengewehre.

Appendix 3

Structure of an Armored Scout Platoon
Police Action Staff Southeast, 1/21/1942

Zugtrupp
Kopfstärke: 1 Unterführer, 2 Kraftfahrer
Waffen: 3 Pistolen, 2 Karabiner

Troß
Kopfstärke: 4 Männer, 4 Kraftfahrer
Waffen: 8 Pistolen, 6 Karabiner, 2 l.MG.

Pkw. 4   Krad

Mlkw. 23   Pzkw. Tatra VII (Reserve)

Panzerspähtrupp I
Kopfstärke: 2 Unterführer, 6 Männer, 6 Kraftfahrer
Waffen: 14 Pistolen, 1 Masch. Pistole, 10 Karabiner, 3 l.MG.

Krad Bwg.   Krad Bwg.   Krad Bwg.   Krad   Krad   Pzkw. Tatra I

Panzerspähtrupp III
Kopfstärke: 2 Unterführer, 6 Männer, 6 Kraftfahrer
Waffen: 14 Pistolen, 1 Masch. Pistole, 10 Karabiner, 3 l.MG.

Krad Bwg.   Krad Bwg.   Krad Bwg.   Krad   Krad   Pzkw. Tatra III

Panzerspähtrupp V
Kopfstärke: 2 Unterführer, 6 Männer, 6 Kraftfahrer
Waffen: 14 Pistolen, 1 Masch. Pistole, 10 Karabiner, 3 l.MG.

Krad Bwg.   Krad Bwg.   Krad Bwg.   Krad   Krad   Pzkw. Tatra V

Appendix 4

Strength & Equipment Chart
Police Armored Unit (Pol. Pzkw. Abt.), 5/1/1942

Gruppe Führer

Kopfstärke: 1 Offizier, 5 Kraftfahrer

Waffen: 6 Pistolen

Stkw. 4   Krad Bwg.   Krad Bwg.   Krad   Krad

Nachschubkolonne (Gefechts- und Gepäcktroß)

Kopfstärke: 9 Unterführer, 1 Mann, 11 Kraftfahrer

Waffen: 21 Pistolen

Stkw. 4   Stkw. 14   Wkw. mit Feldküche   Wkw. für Verpflegung

Wkw. für Munition   Wkw. für Munition   Wkw. für Munition

Wkw. für Betriebsstoff   Wkw. für Betriebsstoff   Wkw. für Betriebsstoff   Wkw. für Gepäck

Werkstattzug

Kopfstärke: 1 Unterführer, 12 Kraftfahrer

Waffen: 13 Pistolen

Stkw. 8   Krad Bwg.   Wkw. für Reparaturteile   Wkw. für Reparaturteile   Wkw. für Reparaturteile

Werkstattkraftwagen   Werkstattkraftwagen   Werkstattkraftwagen

Appendix 4

Strength and Equipment Chart
Police Armored Unit (Pol. Pzkw. Abt.), 5/1/1942

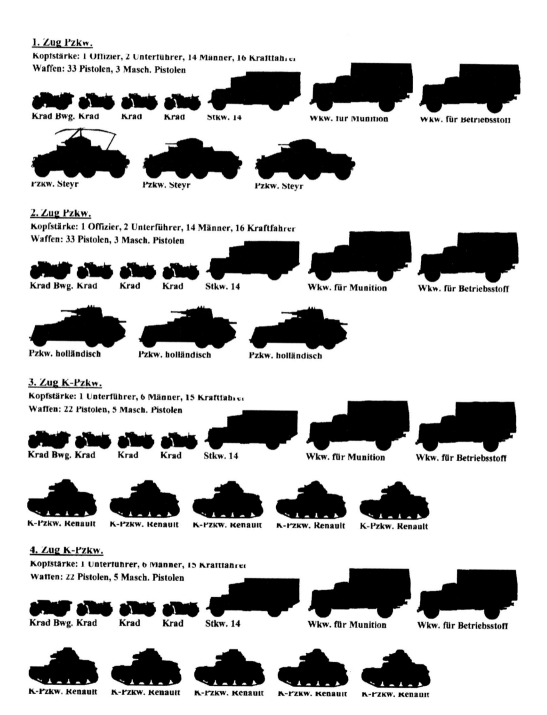

1. Zug Pzkw.

Kopfstärke: 1 Offizier, 2 Unterführer, 14 Männer, 16 Kraftfahrer
Waffen: 33 Pistolen, 3 Masch. Pistolen

Krad Bwg. Krad   Krad   Krad   Stkw. 14   Wkw. für Munition   Wkw. für Betriebsstoff

Pzkw. Steyr   Pzkw. Steyr   Pzkw. Steyr

2. Zug Pzkw.

Kopfstärke: 1 Offizier, 2 Unterführer, 14 Männer, 16 Kraftfahrer
Waffen: 33 Pistolen, 3 Masch. Pistolen

Krad Bwg. Krad   Krad   Krad   Stkw. 14   Wkw. für Munition   Wkw. für Betriebsstoff

Pzkw. holländisch   Pzkw. holländisch   Pzkw. holländisch

3. Zug K-Pzkw.

Kopfstärke: 1 Unterführer, 6 Männer, 15 Kraftfahrer
Waffen: 22 Pistolen, 5 Masch. Pistolen

Krad Bwg. Krad   Krad   Krad   Stkw. 14   Wkw. für Munition   Wkw. für Betriebsstoff

K-Pzkw. Renault   K-Pzkw. Renault   K-Pzkw. Renault   K-Pzkw. Renault   K-Pzkw. Renault

4. Zug K-Pzkw.

Kopfstärke: 1 Unterführer, 6 Männer, 15 Kraftfahrer
Waffen: 22 Pistolen, 5 Masch. Pistolen

Krad Bwg. Krad   Krad   Krad   Stkw. 14   Wkw. für Munition   Wkw. für Betriebsstoff

K-Pzkw. Renault   K-Pzkw. Renault   K-Pzkw. Renault   K-Pzkw. Renault   K-Pzkw. Renault

Appendix 5

Strength and Equipment Chart
Police Armored Company (Pol. Pzkw. Komp.), 7/25/1942

Gruppe Führer
Kopfstärke: 1 Offizier, 4 Kraftfahrer
Waffen: 5 Pistolen, 4 Gewehre

Stkw. 4   Krad Bwg.   Krad   Krad

Nachschubkolonne (Gefechts- und Gepäcktroß)
Kopfstärke: 8 Unterführer, 1 Mann, 8 Kraftfahrer
Waffen: 17 Pistolen, 13 Gewehre

Stkw. 4   Wkw. mit Feldküche   Wkw. für Verpflegung   Wkw. für Gepäck

Wkw. für Munition   Wkw. für Munition   Wkw. für Betriebsstoff   Wkw. für Betriebsstoff

Werkstattzug
Kopfstärke: 1 Unterführer, 7 Kraftfahrer
Waffen: 8 Pistolen, 7 Gewehre

Stkw. 4   Krad Bwg.   Zugwagen mit Tiefladeanhänger

Werkstattkraftwagen   Wkw. für Reparaturteile   Wkw. für Reparaturteile

Appendix 5

Strength and Equipment Chart
Police Armored Company (Pol. Pzkw. Komp.), 7/25/1942

1. Zug Pzkw.
Kopfstärke: 1 Offizier, 2 Unterführer, 14 Männer, 16 Kraftfahrer.
Waffen: 33 Pistolen, 7 Gewehre

2. Zug Pzkw.
Kopfstärke: 1 Offizier, 2 Unterführer, 14 Männer, 16 Kraftfahrer
Waffen: 33 Pistolen, 7 Gewehre

3. Zug K-Pzkw.
Kopfstärke: 1 Unterführer, 6 Männer, 15 Kraftfahrer
Waffen: 22 Pistolen, 7 Gewehre

Appendix 6

Strength Chart
Police Armored Company (t mot.)
(Pol.Panz.Kp. (t mot.)) (13. Pol.-Komp.), 3/27/1943

Gruppe Führer
Kopfstärke: 1 Offizier, 1 Unterführer, 4 Kraftfahrer
Waffen: 6 Pistolen, 4 Gewehre

Stkw. 4   Krad Bwg.   Krad   Krad

Nachschubkolonne (Gefechts- und Gepäcktroß)
Kopfstärke: 9 Unterführer, 4 Männer, 8 Kraftfahrer
Waffen: 19 Pistolen, 16 Gewehre

Stkw. 4   Wkw. mit Feldküche   Wkw. für Verpflegung   Wkw. für Gepäck

Wkw. für Munition   Wkw. für Munition   Wkw. für Betriebsstoff   Wkw. für Betriebsstoff

Werkstattzug
Kopfstärke: 1 Unterführer, 7 Kraftfahrer
Waffen: 8 Pistolen, 7 Gewehre

Stkw. 4   Krad Bwg.   Zugwagen mit Tiefladeanhänger

Werkstattkraftwagen   Wkw. für Reparaturteile   Wkw. für Reparaturteile

Appendix 6

Strength Chart
Police Armored Company (t mot.)
(Pol.Panz.Kp. (t mot)) (13. Pol.-Komp.), 3/27/1943

1. Zug Pzkw.

Kopfstärke: 1 Offizier, 2 Unterführer, 14 Männer, 16 Kraftfahrer
Waffen: 33 Pistolen, 7 Gewehre

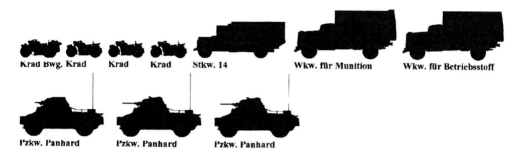

Krad Bwg. Krad Krad Krad Stkw. 14 Wkw. für Munition Wkw. für Betriebsstoff

Pzkw. Panhard Pzkw. Panhard Pzkw. Panhard

2. Zug Pzkw.

Kopfstärke: 1 Offizier, 2 Unterführer, 14 Männer, 16 Kraftfahrer
Waffen: 33 Pistolen, 7 Gewehre

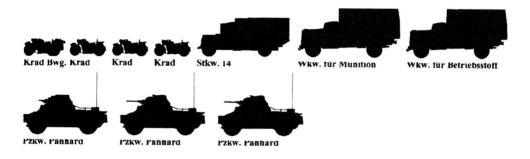

Krad Bwg. Krad Krad Krad Stkw. 14 Wkw. für Munition Wkw. für Betriebsstoff

Pzkw. Panhard Pzkw. Panhard Pzkw. Panhard

3. Zug K-Pzkw.

Kopfstärke: 1 Unterführer, 6 Männer, 15 Kraftfahrer
Waffen: 22 Pistolen, 7 Gewehre

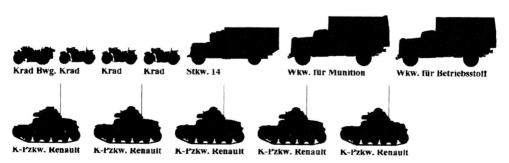

Krad Bwg. Krad Krad Krad Stkw. 14 Wkw. für Munition Wkw. für Betriebsstoff

K-Pzkw. Renault K-Pzkw. Renault K-Pzkw. Renault K-Pzkw. Renault K-Pzkw. Renault

Appendix 7

Strength and Equipment Chart
(strengthened) Police Armored Company (t mot)
(verst.) Pol.Panz.Komp. (t mot), 3/22/1944

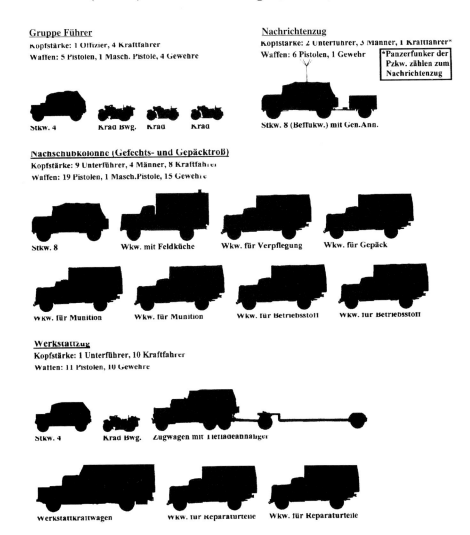

Gruppe Führer

Kopfstärke: 1 Offizier, 4 Kraftfahrer

Waffen: 5 Pistolen, 1 Masch. Pistole, 4 Gewehre

Stkw. 4   Krad Bwg.   Krad   Krad

Nachrichtenzug

Kopfstärke: 2 Unterführer, 3 Männer, 1 Kraftfahrer*

Waffen: 6 Pistolen, 1 Gewehr

*Panzerfunker der Pzkw. zählen zum Nachrichtenzug

Stkw. 8 (Beffukw.) mit Gen.Ann.

Nachschubkolonne (Gefechts- und Gepäcktroß)

Kopfstärke: 9 Unterführer, 4 Männer, 8 Kraftfahrer

Waffen: 19 Pistolen, 1 Masch.Pistole, 15 Gewehre

Stkw. 8   Wkw. mit Feldküche   Wkw. für Verpflegung   Wkw. für Gepäck

Wkw. für Munition   Wkw. für Munition   Wkw. für Betriebsstoff   Wkw. für Betriebsstoff

Werkstattzug

Kopfstärke: 1 Unterführer, 10 Kraftfahrer

Waffen: 11 Pistolen, 10 Gewehre

Stkw. 4   Krad Bwg.   Zugwagen mit Tiefladeanhänger

Werkstattkraftwagen   Wkw. für Reparaturteile   Wkw. für Reparaturteile

Notes:

1. The armament of the armored cars and tanks was listed in special equipment lists according to the type of vehicle.
2. Four men were to be trained as substitute stretcher bearers.
3. The unit formed a gas detection troop, consisting of one leader and three men.
4. Tank radiomen of the 1st platoon were only for armored vehicles with radio sets.
5. The crews of the 2nd, 3rd and 4th platoons operated the radios (when present) themselves.

Appendix 7

Strength and Equipment Chart
(reinforced) Police Armored Company (t mot)
(verst.) Pol.Panz.Komp. (t mot), 3/2/1944

1. Zug Pzkw.
Kopfstärke: 1 Offizier, 5 Unterführer, 17 Männer, 16 Kraftfahrer
Waffen: 35 Pistolen, 9 Gewehre, 1 le. Maschinengewehr

Krad Bwg.   Krad Bwg.   Krad   Krad   Stkw. 14   Wkw. für Munition   Wkw. für Betriebsstoff

Pzkw. Steyr   Pzkw. Steyr   Pzkw. Steyr

2. Zug K-Pzkw
Kopfstärke: 1 Offizier, 5 Unterführer, 5 Männer, 15 Kraftfahrer
Waffen: 24 Pistolen, 9 Gewehre, 1 le. Maschinengewehr

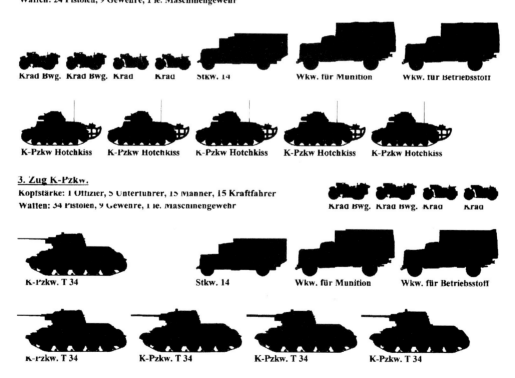

Krad Bwg.   Krad Bwg.   Krad   Krad   Stkw. 14   Wkw. für Munition   Wkw. für Betriebsstoff

K-Pzkw Hotchkiss   K-Pzkw Hotchkiss   K-Pzkw Hotchkiss   K-Pzkw Hotchkiss   K-Pzkw Hotchkiss

3. Zug K-Pzkw.
Kopfstärke: 1 Offizier, 5 Unterführer, 15 Männer, 15 Kraftfahrer
Waffen: 34 Pistolen, 9 Gewehre, 1 le. Maschinengewehr

Krad Bwg.   Krad Bwg.   Krad   Krad

K-Pzkw. T 34   Stkw. 14   Wkw. für Munition   Wkw. für Betriebsstoff

K-Pzkw. T 34   K-Pzkw. T 34   K-Pzkw. T 34   K-Pzkw. T 34

4. Zug K-Pzkw.
wie 3. Zug K-Pzkw.

Appendix 8

Strength and Equipment Chart, 15th Police Armored Company 7/11/1944

(with War Strength Directive (Army) No. 1149, Version B and War Strength Directive (Army) No. 1112 Section b (1st Platoon) added)

Gruppe Führer
Kopfstärke: 1 Offizier, 4 Unterführer, 2 Männer, 4 Kraftfahrer
Waffen: 4 Pistolen, 5 Masch. Pistolen, 2 Gewehre

Wechselbesatzung
Kopfstärke: 1 Untf., 4 Män., 1 Kraftf.
Waffen: 5 Pist., 1 Gew., 1 le. MG.

Stkw. 4   Krad Bwg.   K-Pzkw. M 15   Sturmgeschütz M 42

Wkw. für Mannschaftstransport

Nachschubkolonne (Gefechts- und Gepäcktroß)
Kopfstärke: 8 Unterführer, 10 Männer, 12 Kraftfahrer
Waffen: 4 Pistolen, 2 Masch. Pistolen, 24 Gewehre, 1 le. Maschinengewehr

Stkw. 4   Stkw. 4   Pkw. 4   Wkw. mit Feldküche

Wkw. für Munition   Wkw. für Munition   Wkw. für Munition   Wkw. für Mun. und Gepäck

Wkw. für Betriebsstoff   Wkw. für Betriebsstoff   Wkw. für Betriebsstoff   Wkw. für Gepäck

Instandsetzungsgruppe

a) Werkstattzug
Kopfstärke: 1 Unterführer, 10 Kraftfahrer
Waffen: 11 Pistolen, 10 Gewehre

b) Kompaniekräfte
Kopfstärke: 3 Männer
Waffen: 3 Pistolen, 3 Gewehre

Stkw. 4   Krad Bwg.   Zugwagen mit Tiefladeanhänger

Werkstattkraftwagen   Wkw. für Reparaturteile   Wkw. für Reparaturteile

Appendix 8

Strength and Equipment Chart of the 15th Police Armored Company, 7/11/1944
(with War Strength Directive (Army) No. 1149, Version B, and War Strength Directive (Army) No. 1112, Section b (1st Platoon) added)

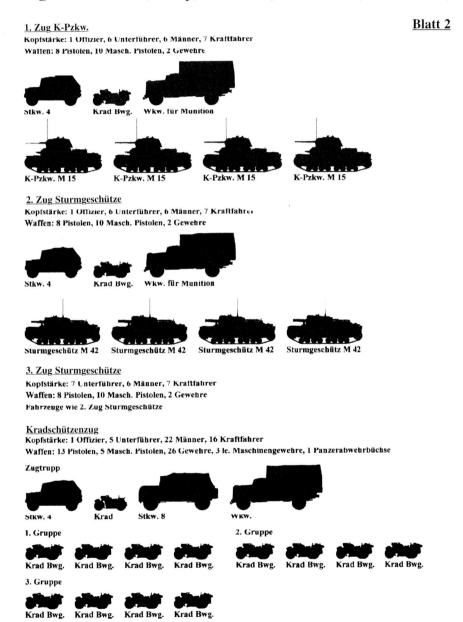

1. Zug K-Pzkw.
Kopfstärke: 1 Offizier, 6 Unterführer, 6 Männer, 7 Kraftfahrer
Waffen: 8 Pistolen, 10 Masch. Pistolen, 2 Gewehre

Stkw. 4   Krad Bwg.   Wkw. für Munition

K-Pzkw. M 15   K-Pzkw. M 15   K-Pzkw. M 15   K-Pzkw. M 15

2. Zug Sturmgeschütze
Kopfstärke: 1 Offizier, 6 Unterführer, 6 Männer, 7 Kraftfahrer
Waffen: 8 Pistolen, 10 Masch. Pistolen, 2 Gewehre

Stkw. 4   Krad Bwg.   Wkw. für Munition

Sturmgeschütz M 42   Sturmgeschütz M 42   Sturmgeschütz M 42   Sturmgeschütz M 42

3. Zug Sturmgeschütze
Kopfstärke: 7 Unterführer, 6 Männer, 7 Kraftfahrer
Waffen: 8 Pistolen, 10 Masch. Pistolen, 2 Gewehre
Fahrzeuge wie 2. Zug Sturmgeschütze

Kradschützenzug
Kopfstärke: 1 Offizier, 5 Unterführer, 22 Männer, 16 Kraftfahrer
Waffen: 13 Pistolen, 5 Masch. Pistolen, 26 Gewehre, 3 le. Maschinengewehre, 1 Panzerabwehrbüchse

Zugtrupp

Stkw. 4   Krad   Stkw. 8   Wkw.

1. Gruppe   2. Gruppe
Krad Bwg. Krad Bwg. Krad Bwg. Krad Bwg.   Krad Bwg. Krad Bwg. Krad Bwg. Krad Bwg.

3. Gruppe
Krad Bwg. Krad Bwg. Krad Bwg. Krad Bwg.

Appendix 9

Strength and Equipment Chart
14th (reinforced) Police Armored Company
(verst.) Pol.-Panz.-Komp., 9/19/1944

Gruppe Führer

Kopfstärke: 1 Offizier, 4 Kraftfahrer

Waffen: 5 Pistolen, 1 Masch. Pistole, 4 Gewehre

Nachrichtenzug

Kopfstärke: 2 Unterführer, 3 Männer, 1 Kraftfahrer*

Waffen: 6 Pistolen, 1 Masch. Pistole, 6 Gewehre

Stkw. 4   Krad Bwg.   Krad   Krad

*Panzerfunker der Pzkw. zählen zum Nachrichtenzug

Stkw. 8 (Beffukw.) mit Gen.Anh.
Behelfsfunkwagen ohne Funkausrüstung

Nachschubkolonne (Gefechts- und Gepäcktroß)

Kopfstärke: 1 Verwaltungsbeamter, 9 Unterführer, 4 Männer, 9 Kraftfahrer

Waffen: 21 Pistolen, 1 Masch. Pistole, 17 Gewehre

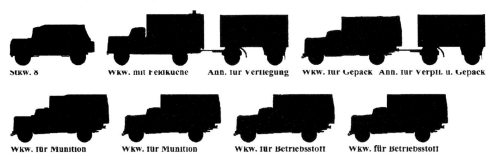

Stkw. 8   Wkw. mit Feldküche   Anh. für Verlegung   Wkw. für Gepäck   Anh. für Verpfl. u. Gepäck

Wkw. für Munition   Wkw. für Munition   Wkw. für Betriebsstoff   Wkw. für Betriebsstoff

Werkstattzug

Kopfstärke: 1 Unterführer, 10 Kraftfahrer

Waffen: 11 Pistolen, 10 Gewehre

Stkw. 4   Krad Bwg.   Zugwagen mit Tiefladeanhänger

Werkstattkraftwagen   Wkw. für Reparaturteile   Wkw. für Reparaturteile

Notes:

1. The armament of the armored vehicles was listed in special equipment lists according to the type of vehicle.
2. Four men were trained as substitute stretcher bearers.
3. The unit formed a gas detection troop of one leader and three men.
4. Tank radiomen were only for vehicles with radios.
5. The crews of the 4th and 5th platoons operated their radios themselves.

Two trailers were assigned as supply-train vehicles; the arrangement here is shown only as an example and was not assigned!

Appendix 9

Strength and Equipment Chart
14th (reinforced) Police Armored Company
(verst.) Pol.-Panz.-Komp., 9/19/1944

1. Zug Pzkw.
Kopfstärke: 1 Offizier, 3 Unterführer, 11 Männer, 16 Kraftfahrer
Waffen: 29 Pistolen, 9 Gewehre, 1 le. Maschinengewehr

Krad Bwg.   Krad Bwg.   Krad   Krad   Stkw. 14   Wkw. (Beute) für Mun.   Wkw. für Betriebsstoff

Pzkw. holländisch   Pzkw. holländisch   Pzkw. holländisch

2. Zug Pzkw.
Kopfstärke: 1 Offizier, 3 Unterführer, 8 Männer, 16 Kraftfahrer
Waffen: 26 Pistolen, 9 Gewehre, 1 le. Maschinengewehr

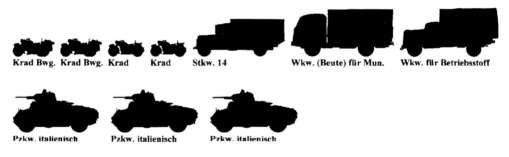

Krad Bwg.   Krad Bwg.   Krad   Krad   Stkw. 14   Wkw. (Beute) für Mun.   Wkw. für Betriebsstoff

Pzkw. italienisch   Pzkw. italienisch   Pzkw. italienisch

3. Zug Pzkw.
Kopfstärke: 1 Offizier, 3 Unterführer, 8 Männer, 13 Kraftfahrer
Waffen: 23 Pistolen, 9 Gewehre, 1 le. Maschinengewehr

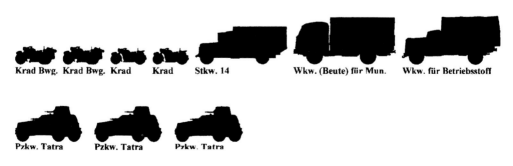

Krad Bwg.   Krad Bwg.   Krad   Krad   Stkw. 14   Wkw. (Beute) für Mun.   Wkw. für Betriebsstoff

Pzkw. Tatra   Pzkw. Tatra   Pzkw. Tatra

Nine trucks for the company were to be taken from captured stocks; the arrangement shown here is only an example!

Appendix 9

Strength and Equipment Chart
14th (reinforced) Police Armored Company
(verst.) Pol.-Pakz.-Komp., 9/19/1944

4. Zug Stu.-Gesch.
Kopfstärke: 1 Offizier, 7 Unterführer, 12 Männer, 15 Kraftfahrer
Waffen: 35 Pistolen, 8 Masch. Pistolen, 3 Gewehre

Sturmgeschütz Sturmgeschütz Stkw. 14 Wkw. (Beute) für Mun. Wkw. (Beute) für Betrst.

Sturmgeschütz Sturmgeschütz Sturmgeschütz Sturmgeschütz Sturmgeschütz Sturmgeschütz

5. Zug Stu.-Gesch.
Kopfstärke: 1 Offizier, 7 Unterführer, 12 Männer, 15 Kraftfahrer
Waffen: 35 Pistolen, 8 Masch. Pistolen, 3 Gewehre
Fahrzeuge wie 4. Zug Stu.-Gesch.

Res. Halbzug Stu.-Gesch.
Kopfstärke: 4 Unterführer, 6 Männer, 6 Kraftfahrer
Waffen: 16 Pistolen, 4 Masch. Pistolen

Sturmgeschütz Sturmgeschütz Sturmgeschütz Sturmgeschütz

Schienenpanzerzug
Kopfstärke: 4 Unterführer, 16 Männer, 14 Kraftfahrer
Waffen: 34 Pistolen, 4 Masch. Pistolen, 2 Gewehre

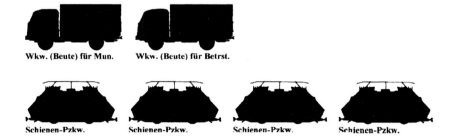

Wkw. (Beute) für Mun. Wkw. (Beute) für Betrst.

Schienen-Pzkw. Schienen-Pzkw. Schienen-Pzkw. Schienen-Pzkw.

Nine trucks for the company were to be taken from captured stocks; the arrangement shown here is only an example!

Sources, Credits and Bibliography

I. Unpublished Sources
Bavarian Main State Archives, Munich
-MInn, Vol. 22   State Ministry of the Interior

Bavarian State Archives, Munich
-Police Administration, Munich

Bavarian State Archives, Nürnberg
-Rep. 218/II   Police Administration, Nürnberg-Fürth

Federal Archives, Koblenz
-R 2   Reich Finance Ministry
-R 19   Main Office, Ordnungspolizei/Chief of the Ordnungspolizei
-R 20   War Diaries
-R 58   Reich Security Headquarters
-R 70   Police offices in absorbed and occupied territories

Federal Archives-Military Archives, Freiburg
-RH 10   Inspector-General of the Armored Troops
-RH 22   Commander of Backline Army Territories
-RH 23   Commanders of backline army territories
-RH 24-1   I. Army Corps
-RH 24-8   VIII. Army Corps
-RH 26-221   221st Securing Division

Police President, Dortmund
-VI   Police Archives

Lower Saxony State Archives, Hannover
-Hann. 180 Lüneburg III, XVI. Police Administration

North Rhine-Westphalian Main State Archives, Düsseldorf
-Administration, Aachen

North Rhine-Westphalian Main State Archives, Düsseldorf, Branch Archives, Kalkum Castle
-PA   Personal records
-BR-Pe Personal records

North Rhine-Westphalian State Archives, Münster
-B 406 1 Arnsberg District, Police Department (Ipa)
-B 131 Bo   Police Headquarters, Bochum

Austrian State Archives, Vienna, Archives of the Republic
-AdR 04 Chancellery, Federal police
-AdR 04 Chancellery, Gendarmerie
-AdR 04 Reich Governor, Vienna

Police Leadership Academy, Münster
-Z 4   Police Historical Collection

Rhineland-Palatinate State Archives, Speyer
-H 74   Police Headquarters, Ludwigshafen

State Archives, Bremen
-4.13/1   Senator for the Interior

State Archives, Hamburg
-Police District 1

II. Published Sources

Official Documents

Special Orders for Provisioning, Chief of the Ordnungspolizei, Kdo. I-Ia (Ib), 1942-1944

German Police, Pocket Calendar for the Schutzpolizei of the Reich and Municipalities and the Verwaltungspolizei, 1942 and 1943

Army Printed Instructions (HDV.) 470/5a: Armored Vehicle Training I (MG), Special Vehicle 101 (Sd.Kfz. 101). 9/2/1939

Army Printed Instructions (HDV.) 470/5b: Armored Vehicle Training II (2 cm), Special Vehicle 121 (Sd.Kfz. 121), 12/24/1938

Army Printed Instructions (HDV.) 470/6: The Light Armored Company, 9/2/1940

Army Printed Instructions (HDV.) 470/9: Guidelines for Forming a Light Armored Replacement Company and Intelligence Replacement Platoon of an Armored Replacement Company

War Strength Directive (KStN.)(Army) No. 1149, Version B: Assault Gun Unit (in Tank Destroyer Unit)(10 or 14 guns), Version B (14 guns), 2/1/1944

War Strength Directive (KStN)(Army) No. 1112, Section b (1st Platoon): Motorcycle Rifle Company (KradSchütz.Kp. b), 1st Platoon, 11/1/1941

Motor Vehicle Technical Appendix to Special Regulations for Maintenance, 1942-1944

Ministerial Sheet for Prussian Interior Administration (MBliV.), 1936

Ministerial Sheet of the Reich and Prussian Ministry of the Interior (RMBliV.), 1936-1944

Information Sheet of the Chief of the Ordnungspolizei

Police Service Instructions (PDV.), 41st Edition, Instructions for Command and Use of Police Troops, Publisher of Police Literature, Lübeck 1943

Newspapers and Magazines

Die Deutsche Polizei (The German Police),4th Year 1936 through 13th Year 1945

Technische Mitteilungen für den Nachrichtenverbindungsdienst der Ordnungspolizei, published by the Test Center for Intelligence Communications, 1944

III. Bibliography

Bader, Kurt, Aufbau und Gliederung der Ordnungspolizei, Berlin 1943

Birn, Ruth Bettina, Die Höheren SS- und Polizeiführer, Himmlers Vertreter im Reich und in den besetzten Gebieten, Düsseldorf 1986

Breitenbach, Jochen, Polizei-Sonderwagen, Geschichte und Einsatz, Gross-Umstadt 1990

Doyle, Chamberlain, Jentz, Encyclopedia of German Tanks of World War Two, London 1978

Fangmann, Reifner, Stangborn, Parteisoldaten, Die Hamburger Polizei in 3. Reich, Hamburg 1987

Fischer, Karl, Waffentechnischer Leitfaden für die Ordnungspolizei, 3. verbesserte und ergänzte Auflage, Berlin 1941

— , Waffen- und Schiesstechnischer Leitfaden für die Ordnungspolizei, 5. verbesserte und erweiterte Auflage, Berlin 1944

Franz, Hermann, Gebirgsjäger der Polizei, Podzun 1963

Goldhagen, Daniel Jonah, Hitlers willige Vollstrecker, Berlin 1996

Kohl, Paul, Der Krieg der deutschen Wehrmacht und der Polizei 1941-1944, Frankfurt 1995

Koschorke, Helmut, Jederzeit einsatzbereit, Ein Bildbericht von der neuen deutschen Polizei, Berlin 1939

Lankenau, Heinrich, Polizei im Einsatz während des Krieges 1939-1945 i Rheinland-Westfalen, Bremen 1957

Luipold, Manfred, Der Polizeisonderwagen. Ein Sonderfahrzeug der Bereitschaftspolizei wird 70 Jahre alt. In: Kastanienfest bei der Bereitschaftspolizei Göppingen, Programmheft 1987, Stuttgart 1987

Mehner, Kurt, Die Waffen-SS und Polizei 1939-1945, Norderstedt 1995

Neufeldt, Huck, Tessin, Zur Geschichte der Ordnungspolizei 1936-1945, Koblenz 1957

Oswald, Werner, Die Kraftfahrzeuge der Polizei und des Bundesgrenzschutzes, Stuttgart 1975

Regenberg, Werner, Beitepanzer unterm Balkenkreuz, Panzerspähwagen und gepanzerte Radfahrzeuge, Wölfersheim-Berstadt 1994

— , Beutepanzer unterm Balkenkreuz, Kleinkampfwagen und gepanzerte Vollkettenschlepper, Wölfersheim-Berstadt 1996

Regenberg, Werner und Scheibert, Horst, Beutepanzer unterm Balkenkreuz, Russische Kampfpanzer, Friedberg 1989

— , Beutepanzer unterm Balkenkreuz, Französische Kampfpanzer, Friedberg 1990

— , Beutepanzer unterm Balkenkreuz, Amerikanische und englische Kampfpanzer, Friedberg 1992

Richter, Hans, Einsatz der Polizei, Bei den Polizei-Bataillonen in Ost, Nord und West, Berlin 1941

— , Ordnungspolizei auf den Rollbahnen des Ostens, Berlin 1943

Roden, Hans, Polizei greift ein, Bilddokumente der Schutzpolizei, Leipzig 1934

Schmidt, Gustav, Strassenpanzerwagen-Sonderwagen der Schutzpolizei, Berlin 1925

Spielberger, Walter, Die gepanzerten Radfahrzeuge des deutschen Heeres 1905-1945, Stuttgart 1974

— , Kraftfahrzeuge und Panzer des österreichischen Heeres 1896 bis heute, Stuttgart 1976

— , Die Motorisierung der Deutschen Reichswehr, Stuttgart 1979

— , Beute-Kraftfahrzeuge und -Panzer der deutschen Wehrmacht, Stuttgart 1989

Wilhelm, Friedrich, Dis Polizei im NS-Staat, Paderborn 1937

Wirth, Hans und Gohler, Fritz, Schutzpolizei im Kampfeinsatz, Handbuch der Taktik des Polizeibataillons, Berlin 1942

IV. Photo Credits

Abbreviations

a.A.	alter Art	old type
Abtl.	Abteilung	Unit
Ag K	Amtsgruppe Kraftfahrwesen	Motor Vehicle Office
Ah	Amperestunde	Amperes per hour
AHA	Allgemeines Heeresamt	General Army Office
AK. A.K.	Armee-Korps	Army Corps
Amp.	Ampere	Ampere
Anl.	Anlage	Appendix, addendum
AOK, A.O.K.	Armee-Oberkommando	Army High Command
Artl.	Artillerie	Artillery
Aufkl.	Aufklärung	Reconnaissance
Ausf.	Ausführung	Version
Batt., Battr.	Batterie	Battery
BdO	Befehlshaber der Ordnungspolizei	Commander of the Ordnungspolizei
BefBlO.	Befehlsblatt der Ordnungspolizei	Orders of the Ordnungspolizei
Beffukw.	Befehlsfunkwagen	Command Radio Truck
Beth.rückw.H.Geb,	Befehlshaber im rückwärtigen Heeresgebiet	Commander in Backline Army Territory
Brig.	Brigade	Brigade
Btl., Battl.	Bataillon	Battalion
Chef H Rüst.u.BdE.	Chef der Heeresrüstung und Befehlshaber des Ersatzheeres	Chief of Army Equipment and Commander of Reserve Army
ChefO.	Chef der Ordnungspolizei	Chief of the Ordnungspolizei
d.	der	of the
d.Sch.P., d.Sch.	der Schutzpolizei	of the Schutzpolizei
Div.	Division	Division
(e)	englisch	British
E, Ers.	Ersatz	Replacement
E., Empf.	Empfanger	Receiver
E.W.	Empfanger WechsNrichter	Receiver-vibrator
Erl.	Erlass	Order
F.A.	Feldausbildung	Field training
Fgst.	Fahrgestell	Chassis
Flak	Flugabwehrkanone	Anti-aircraft gun (AA)
frz., (f)	französisch	French
Fs.Erl., FS-Ertl.	Fernschriftlicher Erlass	Teletype order
Fu	Funk	Radio
g	geheim	Secret
G.R.	Grenadier Regiment	Grenadier Regiment
Geb.	Gebirgs	Mountain
gem.	gemischt	Mixed
Gen.d.Pol.	General der Polizei	General of the Police
Gen.Insp.d.Pz.Tr.	General Inspekteur der Panzertruppe	Inspector-General of the Armored Troops
Gen.Kdo.	Generalkommando	General Command
Gend.	Gendarmerie	Gendarmerie, rural police
Gesch.	Geschütz	Gun
Grp., Gr.	Gruppe	Group
gvH	garnisonsverwendungsfähig Heimat	Capable of home garrison duty
H Rüst	Heeresrüstung	Army Equipment
H.Gr.	Heeresgruppe	Army Group
Hdtsch., Hdsch	Hundertschaft	Hundred, unit of 100 men
HDV	Heeres-Dienstvorschrift	Army Service Instructions
HKL	Hauptkampflinie	Main battle line
HM	Heeres-Mitteilungen	Army Information
holl. (h)	holländisch	of the Netherlands
Hptm.	Hauptmann	Captain
HSSPF.	Höherer SS und Polizeiführer	High SS and Police Leader
I.	im	in the
Ia	Erster Generalstabsoffizier (Führungs-Abteilung)	First General Staff Officer (Command Unit)
IMKK	Interalliierte Militär-Kontrollkommission	Inter-allied Military Control Commission
In 6	Inspektion für Kraftfahrwesen	Inspection of Motor Vehicles
Inf.Div., I.D.	Infanterie-Division	Infantry Division
Inf.Rgt., I.R.	Infanterie-Regiment	Infantry Regiment
ital., (I)	italienisch	Italian
Jäg.Brig.	Jäger-Brigade	Mountain Rifle Brigade
k	kurz	Short
K.-Staffel	Kraftfahrstaffel	Motor Vehicle Echelon
K-Pzkw	Panzerkampfwagen (Ketten-Fahrzeug)	Battle Tank (tracked)
KdO.	Kommandeur der Ordnungspolizei	Commander of the Ordnungspolizei
Kdo.	Kommando	Command
Kdo.Pers.	Kommando-Personalamt	Command Personnel Office
Kdr.	Kommandeur	Commander
Kfz.	Kraftfahrzeug	Motor Vehicle
kHz	Kiloherz	Kilohertz
Kol.	Kolonne	Column
Komp., Kp.	Kompanie	Company

286

Korück	Kommandant des rückwa:rtigen Armeegebietes	Commandant of Backline Army Territory
Kpfwg.	Kampfwagen	Combat Vehicle
Kpzkw.	Ketten-Panzerkraftwagen	Tracked Armored Vehicle
Krad	Kraftrad (Motorrad)	Motorcycle
Krad m.b., Bwg.	Kraftrad mit Beiwagen	Motorcycle with sidecar
kroat.	kroatisch	Croatian
Kschtz.	Kradschützen	Motorcicle Riflemen
KStN.	Kriegsstärke-Nachweisung	War Strength Directive
KTB	Kriegstagebuch	War Diary
Kw.	Kraftwagen	Motor vehicle
KwK, Kwk.	Kampfwagenkanone	Tank gun
le., l.	leicht	Light
l.I.G., le.I.G.	leichtes Infanterie-Geschütz	Light Infantry Gun
LaS	landwirtschaftlicher Schlepper	Farm tractor
le.Sp.	leichter Schienenpanzer	Light armored railcar
lett.	lettisch	Latvian
lfd.	laufend	Continuing, current, running
Lkw.	Lastkraftwagen	Truck
lMG	leichtes Maschinengewehr	Light machine gun
Ltn.	Leutnant	Lieutenant
M.	Modell	Model
MBliV.	Ministerialblatt für die innere Verwaltung	Ministerial Sheet for Interior Administration
MdI.	Ministerium des Inneren	Ministry of the Interior
MG	Maschinengewehr	Machine Gun
Mhz.	Megaherz	Megahertz
Mlkw.	Mannschaftstransportlastkraftwagen	Personnel Carrier Truck
MLKw. 23	Mannschaftstransportlastkraftwagen 23- bis 28sitzig	Personnel Carrier Truck with 23 to 28 seats
mot.	motorisiert	Motorized
MP.	Maschinenpistole	Machine pistol
Mw.	Mittelwelle	Medium-wave
n.A.	neuer Art	New type
Nachr.	Nachrichten	News, Intelligence
Nr.	Nummer	Number
NSDAP	Nationalsozialistische Deutsche Arbeiterpartei	National Socialist German Workers' Party, Nazi Party
III S I d 4		(Registry number for Prussian Ministry of the Interior)
NfD	Nur für den Dienstgebrauch	Only for service use
Gr.Org.	Gruppe Organisation	Group Organization
Obltn.	Oberleutnant	First Lieutenant
Offz.	Offizier	Officer
OKH.	Oberkommando des Heeres	Army High Command
Pak.	Panzerabwehrkanone	Antitank Gun
PBklV, Pol.Bekl.V.	Polizei-Bekleidungs-Vorschrift	Police Clothing Regulation
PDV.	Polizei-Dienstvorschrift	Police Service Regulation
Pi.	Pionier	Engineer
Pkw.	Personenkraftwagen	Car, passenger vehicle
Pkw. 4	Personenkraftwagen 4-sitzig	Four-seat passenger car
Pol.	Polizei	Police
Pol.-E.-Stab	Polizei-Einsatz-Stab	Police Action Staff
Pol.F.Btl.	Polizei-Freiwilligen-Bataillon	Police Volunteer Battalion
Pol.Freiw.Rgt.	Polizei-Freiwilligen-Regiment	Police Vonunteer Regiment
Pol.Geb.Jg.Rgt.	Polizei-Gebirgs-Jäger-Regiment	Police Mountain Rifle Regiment
Pol.Präs.	Polizei-Präsident	Police Commissioner
poln., (p)	polnisch	Polish
preuss.	preussisch	Prussian
Pskw.	Polizei Sonderkraftwagen	Special Police Vehicle
PSW.	Polizei-Sonderwagen	Special Police Vehicle
PV, P.V.	Polizeiverwaltung	Police Administration, Headquarters
PWS	Polizei-Waffen-Schule	Police Weapons School
Pz.	Panzer	Armor, armored vehicle
Pz.Jg.	Panzerjäger	Tank destroyer
Pz.Sp.Wg.	Panzer-Späh-Wagen	Armored scout car
Pzkw.	Panzerspähkraftwagen (auf Rädern)	Armored scout car (on wheels)
RAL	Reichsausschuss für Lieferbedingungen	Reich Delivery Service
RdErl.	Runderlass	General Order
Res.	Reserve	Reserve
Res.Div.	Reserve Division	Reserve Division
Rev.Ltn.	Revier-Leutnant	Precinct Lieutenant
Rev.Offz.	Revier-Offizier	Precinct Officer
RFSSuChdDtPol	Reichsführer SS und Chef der deutschen Polizei	SS Reich Leader and Chief of the German Police
Rgt.	Regiment	Regiment
RMBliV.	Ministerialblatt des Reichs- und Preussischen Ministeriums des Inneren	
	Ministerial Sheet of the Reich and Prussian Ministry of the Interior	
RMdI.	Reichsministerium des Inneren	Reich Ministry of the Interior
RuPrMdI.	Reichs- und Preussisches-Ministerium des Inneren	Reich and Prussian Ministry of the Interior
russ., (r)	russisch	Russian, from the Soviet Union
S.	Sender	Transmitter

s., schw.	schwer	Heavy
SA	Sturmabteilung	Storm Unit, storm troopers
SB	Sammelbezeichnung	Collective designation
Schtz.	Schützen	Riflemen
Schutzm.Schm.	Schutzmannschaft	Police unit
Sd.	Sonder	Special
Sd.Kfz.	Sonderkraftfahrzeug	Special Vehicle
Sich.Div.	Sicherungs-Division	Security Division
Sich.Rgt.	Sicherungs-Regiment	Security Regiment
sMG	schweres Maschinengewehr	Heavy Machine Gun
SS	Schutzstaffel	Security Police
SS-Ogrf.	SS-Obergruppenführer	SS Senior Group Leader
SSPF, SSuPolFhr.	SS- und Polizeiführer	SS and Police Leader
SSPGebF.	SS- und Polizei-Gebietsführer	SS and Police District Leader
SSPSttOF	SS- und Polizei-Standortführer	SS and Police Garrison Leader
Sstkw.	Sonder-Streifenkraftwagen	Special Patrol Vehicle
Stkw. 8	Streifenkraftwagen 8- bis 10sitzig	Patrol Vehicle with 8 to 10 seats
Stkw. 4	Streifenkraftwagen 4sitzig	Patrol Vehicle with 4 seats
Stkw. 14	Streifenkraftwagen 14- bis 16sitzig	Patrol Vehicle with 14 to 16 seats
StuG	Sturmgeschütz	Assault Gun
StVZO	Strassen-Verkehrs-Zulassungsordnung	Road Traffic Permit
(t)	tschechisch	Czech, Czechoslovakian
Techn.Pol.Schule	Technische Polizeischule	Police Technical School
tmot, t mot	teilmotorisiert	Partly motorized
TN	Technische Nothilfe	Technical Emergency Service
to	Tonnen	Tons
Torn.	Tornister	Field pack, portable (radio)
Trp.	Trupp	Troop
U.	Umformer	Transformer
Ukr.Schutzm,Btl.	Ukrainisches-Schutzmannschafts-Bataillon	Ukrainian Police Battalion
Ukw.	Ultrakurzwelle	Ultra-short-wave
v.	vom	Of the, from the
v.H.	von Hundert (%)	Percent
verschrott.	verschrottet	Scrapped
verst., vst.	verstärkt	Reinforced, strengthened
VK	Versuchskonstruktion	Experimental design, test design
Vorläuf.	vorläufige	Temporary, for the time being
Vs	Versuchs	Experimental, test
W.	Watt	Watt
W., Wgn.	Wagen	Vehicle, car, truck, wagon
Wachtm.	Wachtmeister	Patrolman, policeman
WH	Wehrmacht-Heer	Armed Forces-Army
WK	Wehrkreis	Military Area
Wkw.	Wirtschaftskraftwagen	Delivery truck, supply truck
z.b.V.	zur besonderen Verwendung	For special use
Zugw.	Zugwagen	Towing tractor, tow truck
ZW	Zugführer Wagen	Platoon leader's vehicle

Registration Symbols of the Ordnungspolizei Command Office

O.Kdo.	Chef der Ordnungspolizei, Kommandoamt	Chief of the Ordnungspolizei, Command Office
Kdo.	Kommandoamt	Command Office
I	Amtsgruppe I	Official Group I
Ia (1)	Untergruppe Aufgaben, Verwendung, Führung und Einsatz	
	Sub-group I, Tasks, Utilization, Command and Action	
Ia	Gruppe Aufgaben, Verwendung, Führung, Einsatz, Nachschub	
	Group for Tasks, Utilization, Command, Action, Supplying	
In-K	Inspektion Kraftfahrwesen	Inspection of Motor Vehicles
K	Gruppe Kraftfahrwesen	Motor Vehicle Group
K (1b)	Gruppe Kraftfahrwesen: Organisation	Motor Vehicle Group: Organization
K (2)	Gruppe Kraftfahrwesen: Ausbildung, Einsatz	Motor Vehicle Group: Training, Action
K (3)	Gruppe Kraftfahrwesen: Ausrüstung	Motor Vehicle Group: Equipping
O	Gruppe Organisation	Organization Group
O (1)	Gruppe Organisation: Verwendung der uniformierten Ordnungspolizei	
	Organization Group: Utilization of Uniformed Ordnungspolizei	
O (3)	Gruppe Organisation: Organisation der Schutzpolizei des Reiches	
	Organization Group; Organization of National Police	
Org./Ia	Gruppe Organisation: Aufgaben, Verwendung, Führung und Einsatz	
	Organization Group: Tasks, Utilization, Command and Action	
T	Technisches Amt	Technical Office
T (2)	Technisches Amt: Funkwesen	Technical Office: Radio Systems
T (G1)	Technisches Amt: Waffen und Geräte, Personalien	Technical Office: Weapons and Equipment, Personal Effects
VuR	Amt Verwaltung und Recht	Administrative and Legal Office